Springer Undergraduate Mathematics Series

Advisory Board

M.A.J. Chaplain, *University of Dundee, Dundee, Scotland, UK*
K. Erdmann, *University of Oxford, Oxford, England, UK*
A. MacIntyre, *Queen Mary, University of London, London, England, UK*
E. Süli, *University of Oxford, Oxford, England, UK*
M.R. Tehranchi, *University of Cambridge, Cambridge, England, UK*
J.F. Toland, *University of Cambridge, Cambridge, England, UK*

More information about this series at http://www.springer.com/series/3423

Franco Vivaldi

Mathematical Writing

Franco Vivaldi
School of Mathematical Sciences
Queen Mary, University of London
London
UK

ISSN 1615-2085 ISSN 2197-4144 (electronic)
ISBN 978-1-4471-6526-2 ISBN 978-1-4471-6527-9 (eBook)
DOI 10.1007/978-1-4471-6527-9

Library of Congress Control Number: 2014944546

Mathematics Subject Classification: 00A05

Springer London Heidelberg New York Dordrecht

© Springer-Verlag London 2014

This work is subject to copyright. All rights are reserved by the Publisher, whether the whole or part of the material is concerned, specifically the rights of translation, reprinting, reuse of illustrations, recitation, broadcasting, reproduction on microfilms or in any other physical way, and transmission or information storage and retrieval, electronic adaptation, computer software, or by similar or dissimilar methodology now known or hereafter developed. Exempted from this legal reservation are brief excerpts in connection with reviews or scholarly analysis or material supplied specifically for the purpose of being entered and executed on a computer system, for exclusive use by the purchaser of the work. Duplication of this publication or parts thereof is permitted only under the provisions of the Copyright Law of the Publisher's location, in its current version, and permission for use must always be obtained from Springer. Permissions for use may be obtained through RightsLink at the Copyright Clearance Center. Violations are liable to prosecution under the respective Copyright Law.
The use of general descriptive names, registered names, trademarks, service marks, etc. in this publication does not imply, even in the absence of a specific statement, that such names are exempt from the relevant protective laws and regulations and therefore free for general use.
While the advice and information in this book are believed to be true and accurate at the date of publication, neither the authors nor the editors nor the publisher can accept any legal responsibility for any errors or omissions that may be made. The publisher makes no warranty, express or implied, with respect to the material contained herein.

Springer is part of Springer Science+Business Media (www.springer.com)

To Mary-Carmen

Preface

This book has evolved from a course in *Mathematical Writing* offered to second-year undergraduate students at Queen Mary, University of London.

Instructions on writing mathematics are normally given to postgraduate students, because they must write research papers and a thesis. However, there are compelling reasons for providing a similar training at undergraduate level, and, more generally, for raising the profile of writing in a mathematics degree.[1]

A researcher knows that writing an article, presenting a result in a seminar, or simply explaining ideas to a colleague are decisive tests of one's understanding of a topic. If a sketched argument has flaws, these flaws will surface as soon as one tries to convince someone else that the argument is correct. The act of exposition is inextricably linked to thinking, understanding, and self-evaluation.

For this reason, undergraduate students should be encouraged to elucidate their thinking in writing, and to assume greater responsibility for the quality of the exposition of their ideas. Their first-written submissions tend to be cryptic collections of symbols, which easily hide from view learning inadequacies and fragility of knowledge. It is quite possible to perform a correct calculation having only limited understanding of the subject matter, but it is not possible to write about it. A good writing assignment exposes bad studying habits (approaching formal concepts informally, or treating them as mere processes—see [1]), and provides a most effective tool for correcting them.

This course's declared objective is to teach the students how to develop and present mathematical arguments, in preparation for writing a thesis in their final year. For those who will not write a thesis, this course represents an indispensable minimal alternative, which is also more manageable in terms of teaching resources. The writing material is taken mostly from introductory courses in calculus and algebra. This suffices to challenge even the most capable students, who commented on the "unexpected depth" required in their thinking, once forced to offer verbal explanations. The students are asked to use words and symbols with the

---

[1] The poor quality of student writing in Higher Education has raised broad concern [44].

same clarity, precision, and conciseness found in books and lecture notes. This demanding exercise encourages logical accuracy, attention to structure, and economy of thought—the attributes of a mathematical mind. It also forces us to understand better the mathematics we are supposed to know.

The development of writing techniques proceeds from the particular to the general, from the small to the large: words, phrases, sentences, paragraphs, to end with short compositions. These may represent the introduction of a concept, the proof of a theorem, the summary of a section of a book, and the first few slides of a presentation.

The first chapter is a warm-up, listing do's and don'ts of writing mathematics. An essential dictionary on sets, functions, sequences, and equations is presented in Chaps. 2 and 3; these words are then used extensively in simple phrases and sentences. The analysis of mathematical sentences begins in Chap. 4, where we develop some constructs of elementary logic (predicates, quantifiers). This material underpins the expansion of the mathematical dictionary in Chap. 5, where basic attributes of real functions are introduced: ordering, symmetry, boundedness, and continuity. Mathematical arguments are studied in detail in the second part of the book. Chapters 7 and 8 are devoted to basic proof techniques, whereas Chap. 9 deals with existence statements and definitions. Some chapters are dedicated explicitly to writing: Chap. 1 gives basic guidelines; Chap. 6 is concerned with mathematical notation and quality of exposition; Chap. 10 is about writing a thesis. Solutions and hints to selected exercises are given in "Solutions to Exercises".

The symbol [∅] appears often in exercises. It indicates that the written material should contain *no mathematical symbols*. (The allied symbol [∅, n] specifies an approximate word length n of the assignment.) In an appropriate context, having to express mathematics without symbols is a very useful exercise. It brings about the discipline needed to use symbols effectively, and is invaluable for learning how to communicate to an audience of non-experts. Consider the following question: [∅, 100].

> *I have a circle and a point outside it, and I must find the lines through this point which are tangent to the circle. What shall I do?*

The mathematics is elementary; yet answering the question requires a clear understanding of the structure of the problem, and a fair deal of organisation.

> *Write down the equation of a line passing through the point. This equation depends on one parameter, the line's slope, which is the quantity to be determined.*

> *Adjoin the equation of the line to that of the circle, and eliminate one of the unknowns. After a substitution, you'll end up with a quadratic equation in one unknown, whose coefficients still depend on the parameter.*

> *Equate the discriminant of the quadratic equation to zero, to obtain an equation—also quadratic—for the slope. Its two solutions are the desired slopes of the tangent lines. Any geometrical configuration involving vertical lines (infinite slope) will require some care.*

The most challenging exercise of this kind is the MICRO-ESSAY, where the synthesis of a mathematical topic has to be performed in a couple of paragraphs, without using any symbols at all. This exercise prepares the students for writing abstracts, a notoriously difficult task.

The available literature on mathematical writing is almost entirely targeted to postgraduate students and researchers. An exception is *How to think like a mathematician*, by K. Houston [20], written for students entering university, which devotes two early chapters to mathematical writing. The advanced texts include *Mathematical writing,* by D.E. Knuth, T.L. Larrabee, and P.M. Roberts [24], *Handbook of writing for the mathematical sciences* by N. Higham [19], and *A primer of mathematical writing* by S.G. Krantz [22]. Equally valuable is the concise classic text *Writing mathematics well,* by L. Gillman [13]. Unfortunately, this 50-page booklet is out of print, and used copies may command high prices.

The timeless, concise book *The elements of style,* by W. Strunk Jr. and E.B. White [37] is an ideal complement to the present textbook. Anyone interested in writing should study it carefully.

This book was inspired by the lecture notes of a course in logic given by Wilfrid Hodges at Queen Mary in 2005–2006. This course used writing as an essential tool, in an innovative way. Wilfrid has been an ideal companion during a decade-long effort to bring writing to centre stage in our mathematics degree, and I am much indebted to him.

The development of the first draft of this book was made possible by a grant from the Thinking Writing Secondment Scheme, at Queen Mary, University of London; their support is gratefully acknowledged. I also thank Will Clavering and Sally Mitchell for their advice and encouragement throughout this project, and Ivan Tomašić for the discussions on logic over coffee. The students of the Mathematical Writing course at Queen Mary endured the many early versions of this work; I thank them for spotting errors and mis-prints. I am very grateful to Lara Alcock for her thoughtful, detailed feedback on an earlier version of the manuscript, which prompted me to re-consider the structure of some chapters.

Finally, I owe a special debt to my daughter Giulia who had the patience to read the final draft, suggesting improvements in hundreds of sentences and removing hundreds of commas.

London, October 2014   Franco Vivaldi

Contents

1	**Some Writing Tips**	1
1.1	Grammar	1
1.2	Numbers and Symbols	3
1.3	Style	5
1.4	Preparation and Structure	6
2	**Essential Dictionary I**	9
2.1	Sets	9
	2.1.1 Defining Sets	13
	2.1.2 Arithmetic	15
	2.1.3 Sets of Numbers	17
	2.1.4 Writing About Sets	19
2.2	Functions	21
2.3	Some Advanced Terms	26
	2.3.1 Families of Sets	26
	2.3.2 Sums and Products of Sets	26
	2.3.3 Representations of Sets	28
3	**Essential Dictionary II**	31
3.1	Sequences	31
3.2	Sums	34
3.3	Equations and Identities	37
3.4	Expressions	40
	3.4.1 Levels of Description	41
	3.4.2 Describing Expressions	43
3.5	Some Advanced Terms	47
	3.5.1 Sets and Sequences	47
	3.5.2 More on Equations	48

4	**Mathematical Sentences**		51
	4.1	Relational Operators	52
	4.2	Logical Operators	53
	4.3	Predicates	57
	4.4	Quantifiers	59
		4.4.1 Quantifiers and Functions	64
		4.4.2 Existence Statements	66
	4.5	Negating Logical Expressions	67
	4.6	Relations	70
5	**Describing Functions**		77
	5.1	Ordering Properties	77
	5.2	Symmetries	78
	5.3	Boundedness	80
	5.4	Neighbourhoods	81
		5.4.1 Neighbourhoods and Sets	82
	5.5	Continuity	83
	5.6	Other Properties	85
	5.7	Describing Sequences	87
6	**Writing Well**		93
	6.1	Choosing Words	93
	6.2	Choosing Symbols	96
	6.3	Improving Formulae	101
	6.4	Writing Definitions	104
	6.5	Introducing a Concept	106
	6.6	Writing a Short Description	108
7	**Forms of Argument**		113
	7.1	Anatomy of a Proof	113
	7.2	Proof by Cases	116
	7.3	Implications	117
		7.3.1 Direct Proof	118
		7.3.2 Proof by Contrapositive	119
	7.4	Conjunctions	121
		7.4.1 Loops of Implications	122
	7.5	Proof by Contradiction	123
	7.6	Counterexamples and Conjectures	124
	7.7	Wrong Arguments	127
		7.7.1 Examples Versus Proofs	127
		7.7.2 Wrong Implications	128
		7.7.3 Mishandling Functions	129
	7.8	Writing a Good Proof	131

8	**Induction**		139
	8.1	The Well-Ordering Principle	140
	8.2	The Infinite Descent Method	141
	8.3	Peano's Induction Principle	143
	8.4	Strong Induction	145
	8.5	Good Manners with Induction Proofs	147
9	**Existence and Definitions**		151
	9.1	Proofs of Existence	151
	9.2	Unique Existence	155
	9.3	Definitions	157
	9.4	Recursive Definitions	158
	9.5	Wrong Definitions	161
10	**Writing a Thesis**		165
	10.1	Theses and Other Publications	165
	10.2	Title	168
	10.3	Abstract	170
	10.4	Citations and Bibliography	174
		10.4.1 Avoiding Plagiarism	177
	10.5	TeX and LaTeX	177
Solutions to Exercises			181
References			197
Index			199

How to Use this Book

Students

This book is appropriate for self-study at undergraduate or master level. It will be of interest not only to dissertation-writing students looking for specific advice, but also to students at earlier stages of a mathematics programme, and for various purposes.

For example, a student may wish to raise the quality of their written output in coursework and exams, or may need to expand their vocabulary to reflect a growing mathematical maturity. Others may be interested in writing about a particular construct, or may seek to improve the form and style of their proofs.

One year of university mathematics should provide sufficient background for reading this book. Parts of it may be read earlier, even at the start of university if one has a good sense of conceptual accuracy. There is a large supply of exercises; solutions and hints are provided to facilitate self-assessment in absence of a teacher.

We explain succinctly the purpose of each chapter:

1. Getting into the spirit of the course.
2. Writing about sets and functions.
3. Writing about sequences and equations.
4. The logic underpinning complex sentences.
5. Advanced writing on functions.
6. Improving various aspects of a mathematical text.
7. Structuring a mathematical proof.
8. Using induction arguments.
9. Writing definitions.
10. Writing a thesis.

Chapter 1 should be read by everyone. Chapters 2 and 3 are fundamental but introductory; an advanced student should be familiar with much of their content, and will consult them only if necessary. Chapter 4 develops the basic logic needed, say, for a first analysis course. This material is a pre-requisite for Chaps. 5, 7, 9, while Chaps. 6, 10, and most of Chap. 8 may be read without it. Chapter 10 requires the mathematical maturity of a final year student. With this in mind, we suggest some possible paths through this book:

- To improve the use of basic terms: Chaps. 2–5.
- To improve conceptual accuracy: Chaps. 4, 6, 7, 9.
- To write better proofs: Chaps. 7, 8.
- To write a dissertation: Chaps. 6, 10.

Teachers

This book has been used for several years as the textbook for a second-year course in mathematical writing at the University of London, in classes of 50–70 students. The syllabus covers Chaps. 1–9, excluding the last section of Chaps. 2–4, which contain specialised material not needed elsewhere. Chapter 1 is assigned as a self-study exercise at the beginning of the course; Chap. 5 is given as a reading assignment during a pause in the teaching mid-way through the course.

The students are given weekly exercises taken from the book. In the early part of the course, the writing is limited to short phrases and sentences. Once the students have consolidated their mathematical vocabulary, more complex assignments are introduced. In a typical one, the students are given a two-page excerpt from a standard first-year textbook introducing a mainstream topic (the logarithm, Euclid's algorithm, complex numbers, etc.). They are asked to write a short document comprising a title, a few concise key points, and a short summary (150–200 words) *without using any mathematical symbol*. This form of writing is demanding but short, hence manageable in large classes. The requirement that no symbols be used has several educational virtues, and the added bonus of making plagiarism more difficult. An example of this assignment is given at the beginning of Sect. 6.6, in the form of a summary of a section of the present book. Related exercises are found at the end of Chap. 6.

The students submit their weekly work electronically, as a single pdf file. They are given complete freedom in the choice of the electronic medium used to generate their files (anything from scanning hand-written pages to LaTeX). This policy works well, and it also encourages independence and sense of responsibility. Each assignment requires only a limited use of symbols, to minimise the lure and distraction of electronic typesetting. Invariably, the best students are keen to learn LaTeX, and they do so without any supervision.

The coursework is marked by postgraduate students, who receive specialised training and are provided with detailed marking schemes. Lecturer and markers meet weekly to fine-tune the marking and resolve unusual cases. Postgraduate students tend to find this experience more instructive than marking conventional exercises.

The coursework constitutes 20–30 % of the assessment for the course, the rest being the final exam. In a small class, it may be desirable to adjust the assessment balance, increasing the weight of coursework and adding a mini-project or a short presentation at the end of the course.

Finally, this book could be used as supplementary material for various courses and programmes: a first-year introduction to mathematical structures, a course in analysis or in logic, a workshop on writing dissertations.

Chapter 1
Some Writing Tips

The following short mathematical sentences are poorly formed in one way or another. Can you identify all the errors and would you know how to fix them? Compare your answers with those given at the end of the book.

Exercise 1.1 Improve the writing.

1. a is positive.
2. Two is the only even prime.
3. If $x > 0\, g(x) \neq 0$.
4. We minus the equation.
5. $x^2 + 1$ has no real solution.
6. When you times it by negative x, $\leqslant$ becomes $\geqslant$.
7. The set of solutions are all odd.
8. $\sin(\pi x) = 0 \Rightarrow x$ is integer.
9. An invertible matrics is when the determinant is non-zero.
10. This infinete sequence has less negative terms.

In this chapter you will learn how to recognise and correct common mistakes, the first step towards writing mathematics well. By the time you reach the exercises at the end of the chapter you should already feel a sense of progress. You should return to this chapter repeatedly, to monitor the assimilation of good practice.

1.1 Grammar

You are advised to use an English dictionary, e.g.,[1] [34], and to recall the basic terminology of grammar (adjective, adverb, noun, pronoun, verb, etc.[2])—see, for instance, [37, pp. 89–95].

---

[1] Abbreviation for the Latin *exempli gratia*, meaning 'for the sake of example'.
[2] Abbreviation for the Latin *et cetera*, meaning 'and the rest'.

© Springer-Verlag London 2014
F. Vivaldi, *Mathematical Writing*, Springer Undergraduate Mathematics Series,
DOI 10.1007/978-1-4471-6527-9_1

- Write in complete sentences. Every sentence should begin with a capital letter, end with a full stop, and contain a subject and a verb. The expression 'A cubic polynomial' is not a sentence because it doesn't have a verb. It would be appropriate as a caption, or a title, but you can't simply insert it in the middle of a paragraph.
- Make sure that the nouns match the verbs grammatically.

  BAD: The set of primes are infinite.
  GOOD: The set of primes is infinite.

  (The verb refers to 'the set', which is singular.)
- Make a pronoun agree with its antecedent.

  BAD: Each function is greater than their minimum.
  GOOD: Each function is greater than its minimum.

  (The pronoun 'its' refers to 'function', which is singular.)
- If possible, do not split infinitives.

  BAD: We have to thoroughly examine this proof.
  GOOD: We have to examine this proof thoroughly.
  BAD: I was taught to always simplify fractions.
  GOOD: I was taught always to simplify fractions.

  (The infinitives are 'to examine' and 'to simplify'.) In some cases a split infinitive may be acceptable, even desirable.

  GOOD: This is a sure way to more than double the length of the manuscript.
- Check the spelling. No point in crafting a document carefully, if you then spoil it with spelling mistakes. If you use a word processor, take advantage of a spell checker. These are some frequent spelling mistakes:

  BAD: auxillary, catagory, consistant, correspondance, impliment, indispensible, ocurrence, preceeding, refering, seperate.

  These are misspelled mathematical words that I have found in mathematics examination papers:

  BAD: arithmatic, arithmatric, derivitive, divisable, falls (false), infinaty, matrics, orthoganal, orthoginal, othogonal, reciprical, scalor, theorom.
- Be careful about distinctions in meaning.
  Do not confuse *it's* (abbreviation for *it is*) with *its* (possessive pronoun).

  BAD: Its an equilateral triangle: it's sides all have the same length.

  Do not confuse the noun *principle* (general law, primary element) with the adjective *principal* (main, first in rank of importance).

  BAD: the principal of induction
  BAD: the principle branch of the logarithm

1.1 Grammar

Do not use *less* (of smaller amount, quantity) when you should be using *fewer* (not as many as).

BAD: There are less primes between 100 and 200 than between 1 and 100.

- Do not use *where* inappropriately. As a relative adverb, *where* stands for *in which* or *to which*; it does not stand for *of which*.

  BAD: We consider the logarithmic function, where the derivative is positive.
  GOOD: We consider the logarithmic function, whose derivative is positive.

  The adverb *when* is subject to similar misuse.

  BAD: A prime number is when there are no proper divisors.
  GOOD: A prime number is an integer with no proper divisors.

- Do not use *which* when you should be using *that*. Even when both words are correct, they have different meanings. The pronoun *that* is defining, it is used to identify an object uniquely, while *which* is non-defining, it adds information to an object already identified.

  The argument that was used above is based on induction.
  [*Specifies which argument.*]
  The following argument, which will be used in subsequent proofs, is based on induction.
  [*Adds a fact about the argument in question.*]

  A simple rule is to use *which* only when it is preceded by a comma or by a preposition, or when it is used interrogatively.

- In presence of parentheses, the punctuation follows strict rules. The punctuation outside parentheses should be correct if the statement in parentheses is removed; the punctuation within parentheses should be correct independently of the outside.

  BAD: This is bad. (Superficially, it looks good).
  GOOD: This is good. (Superficially, it looks like the BAD one.)
  BAD: This is bad, (on two accounts.)
  GOOD: This is good (as you would expect).

1.2 Numbers and Symbols

Effectively combining numbers, symbols, and words is a main theme in this course. We begin to look at some basic conventions.

- A sentence containing numbers and symbols must still be a correct English sentence, including punctuation.

  BAD: $a < b \ a \neq 0$
  GOOD: We have $a < b$ and $a \neq 0$.
  GOOD: We find that $a < b$ and $a \neq 0$.

GOOD:	Let $a < b$, with $a \neq 0$.
BAD:	$x^2 - 7^2 = 0$. $x = \pm 7$.
GOOD:	Let $x^2 - 7^2 = 0$; then $x = \pm 7$.
GOOD:	The equation $x^2 - 7^2 = 0$ has two solutions: $x = \pm 7$.

- Omit unnecessary symbols.

BAD:	Every differentiable real function f is continuous.
GOOD:	Every differentiable real function is continuous.

- If you use small numbers for counting, write them out in full; if you refer to specific numbers, use numerals.

BAD:	The equation has 4 solutions.
GOOD:	The equation has four solutions.
GOOD:	The equation has 127 solutions.
BAD:	Both three and five are prime numbers.
GOOD:	Both 3 and 5 are prime numbers.

- If at all possible, do not begin a sentence with a numeral or a symbol.

BAD:	ρ is a rational number with odd denominator.
GOOD:	The rational number ρ has odd denominator.

- Do not combine operators ($+$, $\neq$, $<$, etc.) with words.

BAD:	The difference $b - a$ is < 0
GOOD:	The difference $b - a$ is negative.

- Do not misuse the implication operator $\Rightarrow$ or the symbol $\therefore$. The former is employed only in symbolic sentences (Sect. 4.2); the latter is not used in higher mathematics.

BAD:	a is an integer $\Rightarrow a$ is a rational number.
GOOD:	If a is an integer, then a is a rational number.
BAD:	$\Rightarrow \ x = 3$.
BAD:	$\therefore \ x = 3$.
GOOD:	hence $x = 3$.
GOOD:	and therefore $x = 3$.

- Within a sentence, adjacent formulae or symbols must be separated by words.

BAD:	Consider $A_n, n < 5$.
GOOD:	Consider A_n, where $n < 5$.
BAD:	Add p k times to c.
BAD:	Add p to c k times.
GOOD:	Add p to c, repeating this process k times.

1.2 Numbers and Symbols

For displayed equations the rules are a bit different, because the spacing between symbols becomes a syntactic element. Thus an expression of the type

$$A_n = B_n, \quad n < 5$$

is quite acceptable (see Sect. 6.3).

1.3 Style

A good sentence needs a lot more than grammatical correctness.

- Give priority to clarity over fancy language. Avoid long and involved sentences; break long sentences into shorter ones.

    BAD: We note the fact that the polynomial $2x^2 - x - 1$ has the coefficient of the x^2 term positive.
    GOOD: The leading coefficient of the polynomial $2x^2 - x - 1$ is positive.
    BAD: The inverse of the matrix A requires the determinant of A to be non-zero in order to exist, but the matrix A has zero determinant, and so its inverse does not exist.
    GOOD: The matrix A has zero determinant, hence it has no inverse.

- Place important words in a prominent position within the sentence. Suppose you are introducing the logarithm:

    BAD: An important example of a transcendental function is the logarithm.

    In this classic bad opening 'an example of something is something else', the focus of attention is the transcendental functions, not the logarithm.

    GOOD: The logarithm is an important example of a transcendental function.
    GOOD: Let us now define a key transcendental function: the logarithm.

    Suppose you wish to emphasise the scalar product:

    BAD: A commonly used method to check the orthogonality of two vectors is to verify that their scalar product is zero.
    GOOD: If the scalar product of two vectors is zero, then the vectors are orthogonal.

- Prefer the active to the passive voice.

    BAD: The convergence of the above series will now be established.
    GOOD: We establish the convergence of the above series.

- Vary the choice of words to avoid repetition and monotony.[3]

---

[3] If necessary, consult a thesaurus.

-

    BAD: The function defined above is a function of both x and y.
    GOOD: The function defined above depends on both x and y.

- Do not use unfamiliar words unless you know their exact meaning.

    BAD: A simplistic argument shows that our polynomial is irreducible.
    GOOD: A simple argument shows that our polynomial is irreducible.

- Do not use vague, general statements to lend credibility to your writing. Avoid emphatic statements.

    BAD: Differential equations are extremely important in modern mathematics.
    BAD: The proof is very easy, as it makes an elementary use of the triangle inequality.
    GOOD: The proof uses the triangle inequality.

- Do not use jargon, or informal abbreviations: it looks immature rather than 'cool'.

    BAD: Spse U subs x into T eq. Wot R T soltns?

- Enclose side remarks within commas, which is very effective, or parentheses (getting them out of the way). To isolate a phrase, use hyphenation—it really sticks out—or, if you have a word processor, *change font* (**but** `don't` <u>overdo</u> it).
- Take punctuation seriously. To improve it, read [40] or [41].

1.4 Preparation and Structure

There are things one must keep in mind when preparing any document.

- Begin by writing your document in draft form, or at least write down a list of key points. Few people are able to produce good writing at the first attempt.
- Consider the background of your readers; are they familiar with the meaning of the words you use? It's easy to write a mathematics text that's too difficult; it's almost impossible to write one that's too easy.
- Form each sentence in your head before writing it down. Then read carefully what you have written. Read it aloud: how does it sound? Have you written what you intended to write? Is it clear? Don't hesitate to rewrite.
- Split the text into paragraphs. Each paragraph should be about one 'idea', and it should be clear how you are moving from one idea to the next. Be prepared to re-arrange paragraphs. The first idea you thought of may not have been the best one; the sequence of arguments you have chosen may not be optimal.
- When you finish writing, consider the opening and closing sentences of your document. The former should motivate the readers to keep reading, the latter should mark a resting place, like the final bars in a piece of music.

1.4 Preparation and Structure

- Word processing has changed the way we write, and often a document is the end-product of several successive approximations. After prolonged editing, one stops seeing things. If you have time, leave your document to rest for a day or two, and then read it again.

Exercise 1.2 Improve the writing, following the guidelines given in this chapter.

1. There are 3 special cases.
2. X is a finite set.
3. It does not tend to infinaty.
4. It follows $x - 1 = y^4$.
5. $\therefore c^{-1}$ is undefined.
6. The product of 2 negatives is positive.
7. We square the equation.
8. We have less solutions than we had before.
9. $x^2 = y^2$ are two othogonal lines.
10. Let us device a strategy for a proof.
11. This set of matrixs are all invertible.
12. If the integral $= 0$ the function is undefined.
13. Purely imaginary is when the real part is zero.
14. Construct the set of vertex of triangles.
15. From the fact that $x = 0$, I can't divide by x.
16. A circle is when major and minor axis are the same.
17. The function f is not discontinuous.
18. Plug-in that expression in the other equation.
19. I found less solutions than I expected.
20. When the discriminant is < 0, you get complex.
21. We prove Euler theorom.
22. The definate integral is where you don't have integration limits.
23. The asyntotes of this hiperbola are othogonal.
24. A quadratic function has 1 stationery point.
25. The solution is not independent of s.
26. a is negative $\therefore \sqrt{a}$ is complex.
27. Thus $x = a$. (We assume that a is positive).
28. Each value is greater than their reciprical.
29. Remember to always check the sign.
30. Differentiate f n times.

Chapter 2
Essential Dictionary I

In writing mathematics we use words and symbols to describe facts. We need to explain the meanings of words and symbols, and to state and prove the facts.

We'll be concerned with facts later. In this chapter and the next we list mathematical words with accompanying notation. This is our essential mathematical dictionary. It contains some 200 entries, organised around few fundamental terms: **set**, **function**, **sequence**, **equation**. As we introduce new words, we use them in short phrases and sentences.

Dictionaries are not meant to be read through, so don't be surprised if you find the exposition demanding. Take it in small doses. The last section of this chapter deals with advanced terminology and may be skipped on first reading.

2.1 Sets

A **set** is a collection of *well-defined*, *unordered*, *distinct* objects. (This is the so-called 'naive definition' of a set, due to Cantor.[1]) These objects are called the **elements** of a set, and a set is determined by its elements. We may write

The set of all odd integers
The set of vertices of a pentagon
The set of differentiable real functions

In simple cases, a set can be defined by listing its elements, separated by commas, enclosed within curly brackets. The expression

$$\{1, 2, 3\}$$

---

[1] Georg Cantor (German: 1845–1918).

denotes the set whose elements are the integers 1, 2 and 3. Two sets are equal if they have the same elements:
$$\{1, 2, 3\} = \{3, 2, 1\}.$$

(By definition, the order in which the elements of a set are listed is irrelevant.)

It is customary to ignore repeated set elements: $\{2, 1, 3, 1, 3\} = \{2, 1, 3\}$. This convention, adopted by computer algebra systems, simplifies the definition of sets. If repeated elements are allowed and not collapsed, then we speak of a **multiset**: $\{2, 1, 3, 1, 3\}$. The **multiplicity** of an element of a multiset is the number of times the element occurs. Reference to multiplicity usually signals that there is a multiset in the background:

Every quadratic equation has two complex solutions, counting multiplicities.

Multisets are not as common as sets.

The set $\{\}$ with no elements is called the **empty set**, denoted by the symbol ∅. The empty set is distinct from 'nothing', it is more like an empty container. For example, the statements

This equation has no solutions.
The solution set of this equation is empty.

have the same meaning.

To assign a symbol to a mathematical object, we use an **assignment statement** (or **definition**), which has the following syntax:

$$A := \{1, 2, 3\}. \tag{2.1}$$

This expression assigns the symbolic name A to the set $\{1, 2, 3\}$, and now we may use the former in place of the latter. The symbol ':=' denotes the **assignment operator**. It reads '*becomes*', or '*is defined to be*', rather than '*is equal to*', to underline the difference between assignment and equality (in computer algebra, the symbols = and := are not interchangeable at all!). So we can't write $\{1, 2, 3\} := A$, because the left operand of an assignment operator must be a symbol or a symbolic expression.

The right-hand side of an assignment statement such as (2.1) is a collection of symbols or words that pick out a unique thing, which logicians call the *definiens* (Latin for 'thing that defines'). The left-hand side is a symbol that will be used to stand for this unique thing, which is called the *definiendum* (Latin for 'thing to be defined'). These terms are rather heavy, but they are widely used [36, Chap. 8]. The definiendum may also be a symbolic expression—see below.

While it's very common to use the equal sign '=' also for an assignment, the specialised notation := improves clarity. There are other symbols for the assignment operator, namely

$$\stackrel{\text{def}}{=} \quad \stackrel{\triangledown}{=}, \tag{2.2}$$

which make an even stronger point.

2.1 Sets

To indicate that x is an element of a set A, we write

$$x \in A \qquad x \text{ is an element of } A \qquad x \text{ belongs to } A.$$

The symbol $\notin$ is used to negate membership. Thus

$$\{7, 5\} \in \{5, \{5, 7\}\} \qquad 7 \notin \{5, \{5, 7\}\}.$$

(Think about it.)

A **subset** B of a set A is a set whose elements all belong to A. We write

$$B \subset A \qquad B \text{ is a subset of } A \qquad B \text{ is contained in } A$$

and we use $\not\subset$ to negate set inclusion. For example

$$\{3, 1\} \subset \{1, 2, 3\} \qquad \emptyset \subset \{1\} \qquad \{2, 3\} \not\subset \{2, \{2, 3\}\}.$$

Every set has at least two subsets: itself and the empty set. Sometimes these are referred to as the **trivial** subsets. Every other subset—if any—is called a **proper subset**. Motivated by an analogy with $\leq$ and $<$, some authors write $\subseteq$ in place of $\subset$, reserving the latter for proper inclusion: $\mathbb{R} \subseteq \mathbb{R}$, $\mathbb{Q} \subset \mathbb{R}$. Proper inclusion is occasionally expressed with the pedantic notation $\subsetneq$.

The **cardinality** of a set is the number of its elements, denoted by the prefix #:

$$\#\{7, -1, 0\} = 3 \qquad \#A = n.$$

The absolute value symbol $|\cdot|$ is also used to denote cardinality: $|A| = n$. Common sense will tell when this choice is sensible. A set is **finite** if its cardinality is an integer, and **infinite** otherwise. To indicate that the set A is finite, without disclosing its cardinality, we write

$$\#A < \infty. \tag{2.3}$$

A more rigorous account of cardinality will be given in Sect. 2.3.3.

Next we consider the words associated with operations between sets. We write $A \cap B$ for the **intersection** of the sets A and B: this is the set comprising elements that belong to both A and B. If $A \cap B = \emptyset$, we say that A and B are **disjoint**, or have **empty intersection**. The sets $A_1, A_2, \ldots$ are **pairwise disjoint** if $A_i \cap A_j = \emptyset$ whenever $i \neq j$.

We write $A \cup B$ for the **union** of A and B, which is the set comprising elements that belong to A or to B (or to both A and B).

We write $A \setminus B$ for the **(set) difference** of A and B, which is the collection of the elements of A that do not belong to B. The **symmetric difference** of A and B, denoted by $A \triangle B$, is defined as

$$A \triangle B \stackrel{\text{def}}{=} (A \setminus B) \cup (B \setminus A).$$

The assignment operator '$\stackrel{\text{def}}{=}$' [cf. (2.2)] makes it clear that this is a definition. This notation establishes the meaning of $A \triangle B$, which is a symbolic expression rather than an individual symbol. The following examples illustrate the action of set operators:

$$\{1, 2, 3\} \cap \{3, 4, 5\} = \{3\}$$
$$\{1, 2, 3\} \cup \{3, 4, 5\} = \{1, 2, 3, 4, 5\}$$
$$\{1, 2, 3\} \setminus \{3, 4, 5\} = \{1, 2\}$$
$$\{1, 2, 3\} \triangle \{3, 4, 5\} = \{1, 2, 4, 5\}.$$

The above **set operators** are **binary**; they have two sets as **operands**. The identities

$$A \cap B = B \cap A \qquad (A \cap B) \cap C = A \cap (B \cap C)$$

express the **commutative** and **associative** properties of the intersection operator. Union and symmetric difference enjoy the same properties, but set difference doesn't.

Let A be a subset of a set X. The **complement** of A (in X) is the set $X \setminus A$, denoted by A' or by A^c. The complement of a set is defined with respect to an **ambient set** X. Reference to the ambient set may be omitted if there is no ambiguity. So we write

The odd integers is the complement of the even integers

since it's clear that the ambient set is the integers.

With set operators we can construct new sets from old ones, although, in a sense, we are recycling things we already have. To create genuinely new sets, we introduce the notion of **ordered pair**. This is an expression of the type (a, b), with a and b arbitrary quantities. Ordered pairs are defined by the property

$$(a, b) = (c, d) \quad \text{if} \quad a = c \text{ and } b = d. \tag{2.4}$$

The ordered pair (a, b) should not be confused with the set $\{a, b\}$, since for pairs order is essential and repetition is allowed. (Ordered pairs may be defined solely in terms of sets—see Exercise 2.14.) Let A and B be sets. We consider the set of all ordered pairs (a, b), with a in A and b in B. This set is called the **cartesian product** of A and B, and is written as

$$A \times B.$$

Note that A and B need not be distinct; one may write A^2 for $A \times A$, A^3 for $A \times A \times A$, etc. Because the cartesian product is **associative**, the product of more than two sets is defined unambiguously. Also note that the explicit presence of the multiplication operator '$\times$' is needed here, because the expression AB has a different meaning [see Eq. (2.21), Sect. 2.3].

2.1 Sets

2.1.1 Defining Sets

Defining a set by listing its elements is inadequate for all but the simplest situations. How do we define large or infinite sets? A simple device is to use the **ellipsis** '...', which indicates the deliberate omission of certain elements, the identity of which is made clear by the context. For example, the set $\mathbb{N}$ of **natural numbers** is defined as

$$\mathbb{N} := \{1, 2, 3, \ldots\}.$$

Here the ellipsis represents all the integers greater than 3. Some authors regard 0 as a natural number, so the definition

$$\mathbb{N} := \{0, 1, 2, 3, \ldots\}$$

is also found in the literature. Both definitions have merits and drawbacks; mathematicians occasionally argue about it, but this issue will never be resolved. So, when using the symbol $\mathbb{N}$, one may need to clarify which version of this set is employed.[2] The set of **integers**, denoted by $\mathbb{Z}$ (from the German *Zahlen*, meaning numbers), can also be defined using ellipses:

$$\mathbb{Z} := \{\ldots, -2, -1, 0, 1, 2, \ldots\} \quad \text{or} \quad \mathbb{Z} := \{0, \pm 1, \pm 2, \ldots\}.$$

To define general sets we need more powerful constructs. A **standard definition** of a set is an expression of the type

$$\{x : x \text{ has } \mathscr{P}\} \tag{2.5}$$

where $\mathscr{P}$ is some unambiguous property that things either have or don't have. This expression identifies the set of all objects x that have property $\mathscr{P}$. The colon ':' separates out the object's symbolic name from its defining properties. The vertical bar '|' or the semicolon ';' may be used for the same purpose.

Thus the empty set may be defined symbolically as

$$\emptyset \stackrel{\text{def}}{=} \{x : x \neq x\}. \tag{2.6}$$

The property $\mathscr{P}$ is 'x is not equal to x', which is not satisfied by any x. Likewise, the cartesian product $A \times B$ of two sets (see Sect. 2.1) may be specified as

$$\{x : x = (a, b) \text{ for some } a \in A \text{ and } b \in B\}.$$

The rule 'x has property $\mathscr{P}$' now reads: 'x is of the form (a, b) with $a \in A$ and $b \in B$'. The same set may be defined more concisely as

---

[2] Some authors denote the second version by the symbol $\mathbb{N}_0$.

$$\{(a, b) : a \in A \text{ and } b \in B\}.$$

This is a variant of the standard definition (2.5), where the type of object being considered (ordered pair) is specified at the outset. This form of standard definition can be very effective.

The set $\mathbb{Q}$ of **rational numbers**—ratios of integers with non-zero denominator—is defined as follows:

$$\mathbb{Q} := \{\frac{a}{b} : a \in \mathbb{Z}, \ b \in \mathbb{N}, \ \gcd(a, b) = 1\}. \tag{2.7}$$

The property $\mathscr{P}$ is phrased in such a way as to avoid repetition of elements. This is the so-called **reduced form** of rational numbers. The rational numbers may also be defined abstractly, as infinite sets of equivalent fractions—see Sect. 4.6.

One might think that in the expression for a set we could choose any property $\mathscr{P}$. Unfortunately this doesn't work for a reason known as the *Russell-Zermelo paradox*[3] (1901). Consider the set definition

$$W := \{x : x \notin x\} \tag{2.8}$$

in which $\mathscr{P}$ is the property of being a set that is not a member of itself. The quantity

$$x = \{3, \{3, \{3, \{3\}\}\}\}$$

has property $\mathscr{P}$ and hence belongs to W, whereas

$$x = \{3, \{3, \{3, \{3, \ldots\}\}\}\} \quad \text{or} \quad x = \{3, x\}$$

does not have property $\mathscr{P}$ and hence does not belong to W. (In the above expression, the nested parentheses must match, so the notation $\{3, \{3, \{3, \{3, \ldots\}$ is incorrect.)

Given that W is a set of sets, we ask: does W belong to W? We see that if $W \in W$, then W has property $\mathscr{P}$, that is, $W \notin W$, and vice-versa. Impossible! Thus the standard definition (2.8), so deceptively similar to (2.6), does not actually define any set.

Fortunately, we can define a set in such a way that the definition guarantees the existence of the set. A **Zermelo definition** identifies a set W by describing it as

The set of members of X that have property $\mathscr{P}$

where the **ambient set** X is given beforehand, and $\mathscr{P}$ is a property that the members of X either have or do not have. In symbols, this is written as

---

[3] Bertrand Russell (British: 1872–1970); Ernst Zermelo (German: 1871–1953).

2.1 Sets

$$W := \{x \in X : x \text{ has } \mathscr{P}\}. \tag{2.9}$$

For example, the expression

The set of real numbers strictly between 0 and 1

is a Zermelo definition: the ambient set is the set of real numbers, and we form our set by choosing from it the elements that have the stated property.

Zermelo definitions work because it's a basic principle of mathematics (the so-called *subset axiom*) that for any set X of objects and any property $\mathscr{P}$, there is exactly one set consisting of the objects that are in X and have property $\mathscr{P}$. In Sect. 4.3 we shall see that the definiens of a Zermelo definition—a sentence with a variable x in it—is just a special type of function, called a **predicate**.

Both styles of definitions, standard and Zermelo, are widely used in mathematical writing.

2.1.2 Arithmetic

The notation for arithmetical operations is familiar and established. The **sum** and **difference** of two numbers x and y are always written $x + y$ and $x - y$, respectively. By contrast, their **product** may be written in several equivalent ways:

$$xy \qquad x \cdot y \qquad x \times y, \tag{2.10}$$

and so may their **quotient**:

$$\frac{x}{y} \qquad x/y \qquad x : y.$$

(The notation $x : y$ is used mostly in elementary texts.) Do not confuse the product dot '$\cdot$' with the **decimal point** '.', e.g., $3 \cdot 4 = 12$ and $3.4 = 17/5$.

The **reciprocal** of x, defined for $x \neq 0$, is also written in several ways:

$$\frac{1}{x} \qquad 1/x \qquad x^{-1}$$

while the **opposite** of x is $-x$.

The notation for exponentiation is x^y, where x is the **base** and y the **exponent**. Defining exponentiation for a general exponent is a delicate matter, as it requires the logarithmic and exponential functions. The case of a positive integer exponent is easier, because exponentiation reduces to repeated multiplication:

$$x^n \stackrel{\text{def}}{=} \underbrace{x \cdots x}_{n} \qquad n \geq 1.$$

The assignment operator $\stackrel{\text{def}}{=}$ [see (2.2)] indicates that this is a definition, giving meaning to the expression on the left. The use of the under-brace is necessary to specify the number of terms in the product, because all terms are identical. Also note the use of the **raised ellipsis** '$\cdots$' to represent repeated multiplication (or repeated applications of any operator), to be compared with the ordinary ellipsis '$\ldots$', used for sets and sequences (see Sect. 3.1). Thus

$$\underbrace{x \cdots x}_{4} = x \cdot x \cdot x \cdot x \qquad \underbrace{x, \ldots, x}_{4} = x, x, x, x$$

whereas the notation $x \ldots x$ is incorrect.

In integer arithmetic, the symbol '|' is used for **divisibility**.

$$3 | x \qquad 3 \text{ divides } x \qquad x \text{ is a multiple of } 3.$$

EXAMPLE. Turn symbols into words:

$$\{x \in \mathbb{Z} : x \geq 0,\ 2 \mid x\}.$$

BAD: The set of integers that are greater than or equal to zero, and such that 2 divides them. (*Robotic.*)
GOOD: The set of non-negative even integers.

A positive divisor of n, which is not 1 or n, is called a **proper divisor**, and a **prime** is an integer greater than 1 that has no proper divisors. The acronyms **gcd** and **lcm** are used for **greatest common divisor** and **least common multiple**. (The expression **highest common factor** (hcf)—a variant of gcd which is popular in schools—is seldom used in higher mathematics.) Two integers are **co-prime** (or **relatively prime**) if their greatest common divisor is 1. Some authors use (a, b) for $\gcd(a, b)$; this is to be avoided, since this notation is already overloaded.

The following concise notation represents certain infinite sets of integers (here k and m are any integers):

$$k + m\mathbb{Z} \stackrel{\nabla}{=} \{x \in \mathbb{Z} : m | (x - k)\}$$
$$= \{\ldots, k - 2m, k - m, k, k + m, k + 2m, \ldots\}. \qquad (2.11)$$

This definition gives meaning to the symbolic expression $k + m\mathbb{Z}$ on the left of the assignment operator, which otherwise would be meaningless (you can't form the sum or product of an integer and a set!). The two expressions on the right represent the same object. While any of the two would suffice, their combination adds clarity (we shall expand this idea in Sect. 6.4).

2.1 Sets

This notation is economical and effective:

$$x \in 1 + 2\mathbb{Z} \qquad x \text{ is odd}$$
$$x \in m\mathbb{Z} \qquad x \text{ is a multiple of } m$$
$$x \in n^2\mathbb{Z} \setminus 2\mathbb{Z} \qquad x \text{ is an odd multiple of } n^2.$$

This is a special case of a more general notation for sums and products of sets, to be developed in Sect. 2.3.

2.1.3 Sets of Numbers

The 'open face' symbols $\mathbb{N}$, $\mathbb{Z}$, $\mathbb{Q}$ were introduced in Sect. 2.1.1 to represent the natural numbers, the integers, and the rationals, respectively. Likewise, we denote by $\mathbb{R}$ the set of **real** numbers (its symbolic definition is left as Exercise 2.13), while the set of **complex** numbers is denoted by $\mathbb{C}$. The set $\mathbb{C}$ may be written as

$$\mathbb{C} \stackrel{\text{def}}{=} \{x + iy : i^2 = -1, \ x, y \in \mathbb{R}\}.$$

The symbol i is called the **imaginary unit**, while x and y are, respectively, the **real part** $\text{Re}(z)$ and the **imaginary part** $\text{Im}(z)$ of the complex number $z = x + iy$. The sets $\mathbb{R}$ and $\mathbb{C}$ are represented geometrically as the **real line** and the **complex plane** (or **Argand plane**), respectively. A plot of complex numbers in the Argand plane is called an **Argand diagram**. We have the chain of proper inclusions

$$\mathbb{N} \subset \mathbb{Z} \subset \mathbb{Q} \subset \mathbb{R} \subset \mathbb{C}.$$

We now construct new sets from the sets of numbers introduced above. An **interval** is a subset of $\mathbb{R}$ of the type

$$[a, b] := \{x \in \mathbb{R} : a \leqslant x \leqslant b\}$$

where a, b are real numbers, with $a < b$. This interval is **closed**, that is, it contains its end points. (A point is sometimes regarded as a degenerate closed interval, by allowing $a = b$ in the definition.) We also have **open** intervals

$$(a, b) := \{x \in \mathbb{R} : a < x < b\}$$

as well as **half-open** intervals

$$[a, b) \qquad (a, b].$$

The notational clash between an open interval $(a, b) \subset \mathbb{R}$ and an ordered pair $(a, b) \in \mathbb{R}^2$ is unfortunate but unavoidable, since both notations are firmly established. For (half) open intervals, there is the following alternative—and very logical—notation

$$]a, b[\qquad [a, b[\qquad]a, b],$$

which, for some reason, is not so common.

The interval with end-points $a = 0$ and $b = 1$ is the (open, closed, half-open) **unit interval**. A semi-infinite interval

$$\{x \in \mathbb{R} : a < x\} \qquad \{x \in \mathbb{R} : x \leqslant b\}$$

is called a **ray**. The rays consisting of all positive real or rational numbers are particularly important, and have a dedicated notation

$$\mathbb{R}^+ := \{x \in \mathbb{R}, \ x > 0\} \qquad \mathbb{Q}^+ := \{x \in \mathbb{Q}, \ x > 0\} \tag{2.12}$$

whereas $\mathbb{Z}^+$ is just $\mathbb{N}$.

Some authors extend the meaning of interval to include also rays and lines, and use expressions such as

$$(-\infty, \infty) \qquad [a, \infty) \qquad (-\infty, b]. \tag{2.13}$$

As infinity does not belong to the set of real numbers, the notation $[1, \infty]$ is incorrect.

A variant of (2.12) is used to denote non-zero real and rational numbers:

$$\mathbb{R}^* := \{x \in \mathbb{R}, \ x \neq 0\} \qquad \mathbb{Q}^* := \{x \in \mathbb{Q}, \ x \neq 0\}. \tag{2.14}$$

This notation is common but not universally recognised; before using these symbols, a clarifying comment may be appropriate (see Sect. 6.2).

The set $\mathbb{R}^2$ of all ordered pairs of real numbers is called the **cartesian plane**, which is the cartesian product of the real line with itself. If $(x, y) \in \mathbb{R}^2$, then the first component x is called the **abscissa** and the second component y the **ordinate**.

The set $\mathbb{Q}^2 \subset \mathbb{R}^2$, the collection of points of the plane having rational coordinates, is called the set of **rational points** in $\mathbb{R}^2$. The set $[0, 1]^2 \subset \mathbb{R}^2$ is called the **unit square**. In $\mathbb{R}^3$ we have the **unit cube** $[0, 1]^3$, and for $n > 3$ we have the **unit hypercube** $[0, 1]^n \subset \mathbb{R}^n$. The following subsets of the cartesian plane are related to the geometrical figure of the circle:

$$\begin{aligned}
\{(x, y) \in \mathbb{R}^2 : x^2 + y^2 = 1\} &\quad \text{unit circle} \\
\{(x, y) \in \mathbb{R}^2 : x^2 + y^2 \leqslant 1\} &\quad \text{closed unit disc} \\
\{(x, y) \in \mathbb{R}^2 : x^2 + y^2 < 1\} &\quad \text{open unit disc.}
\end{aligned} \tag{2.15}$$

Thus the closed unit disc is the union of the open unit disc and the unit circle. The (unit) circle is denoted by the symbol $\mathbb{S}^1$.

2.1 Sets

For $n \geq 0$, the n-**dimensional unit sphere** $\mathbb{S}^n$ is defined as follows:

$$\mathbb{S}^n = \{(x_0, \ldots, x_n) \in \mathbb{R}^{n+1} : x_0^2 + \cdots + x_n^2 = 1\}$$

This Zermelo definition, to be compared with the Definition (2.15) of the unit circle $\mathbb{S}^1$, employs a combination of ordinary and raised ellipses. For $n = 0$, we have $\mathbb{S}^0 = \{-1, 1\}$.

2.1.4 Writing About Sets

The vocabulary on sets developed so far is sufficient for our purpose. We begin to use it in short phrases which define sets.

1. *The set of ordered pairs of complex numbers.*
2. *The set of rational points on the unit circle.*
3. *The set of prime numbers with fifty decimal digits.*
4. *The set of lines in the cartesian plane, passing through the origin.*

Note that we haven't used any symbols. The set in item 1 is $\mathbb{C}^2$. In item 2, among the infinitely many points of the unit circle, we consider those having rational co-ordinates. There is no difficulty in writing this set symbolically:

$$\{(x, y) \in \mathbb{Q}^2 : x^2 + y^2 = 1\}$$

although its properties are not obvious from the definition. This set is non-empty (the points $(0, \pm 1)$, $(\pm 1, 0)$ belong to it), but is it infinite? This example illustrates the power of a verbal definition. Item 3, which defines a subset of $\mathbb{N}$, makes an even stronger point. This set must be extremely large, but can we even show that it is non-empty? In item 4, each line counts as a single element, rather than an infinite collection of points (otherwise our set of lines would be the whole plane). The symbolic definition of this set is awkward; to simplify it we'll consider suitable **representations** of this set (Sect. 2.3.3).

It is possible to specify a *type* of set, without revealing its precise identity. In each of the following sets there is at least one unspecified quantity.

The set of fractions representing a rational number.
The set of divisors of an odd integer.
A proper infinite subset of the unit circle.
The cartesian product of two finite sets of complex numbers.
A finite set of consecutive integers.

Next we define sets in two ways, first with a combination of words and symbols, and then with words only. One should consider the relative merits of the two formulations.

Let $X = \{3\}$.
The set whose only element is the integer 3.
Let $X = \{m\}$, with $m \in \mathbb{Z}$.
A set whose only element is an integer.
Let $m \in \mathbb{Z}$, and let X be a set such that $m \in X$.
A set which contains a given integer.
Let X be a set such that $X \cap \mathbb{Z} \neq \emptyset$.
A set which contains at least one integer.
Let X be a set such that $\#(X \cap \mathbb{Z}) = 1$.
A set which contains precisely one integer.

In the first two examples the combination of 'let' and '=' replaces an assignment operator. An expression such as '*Let $X \stackrel{\nabla}{=} \{3\}$*' would be overloaded.

The distinction between definite and indefinite articles is essential, the former describing a unique object, the latter an unspecified representative of a class of objects. In the following phrases, a change in one article, highlighted in boldface, has resulted in a logical mistake.

BAD: *A proper infinite subset of **a** unit circle.*
BAD: ***A** set whose only element is the integer 3.*
BAD: ***The** set whose only element is an integer.*
BAD: ***The** set which contains precisely one integer.*

As a final exercise, we express some geometric facts using set terminology.

The intersection of a line and a conic section has at most two points.
The set of rational points in any open interval is infinite.
A cylinder is the cartesian product of a segment and a circle.
The complement of the unit circle consists of two disjoint components.

The reader should re-visit familiar mathematics and describe it in the language of sets.

Exercise 2.1 For each of the following topics:

prime numbers, fractions, complex numbers,

(i) write five short sentences; (ii) ask five questions. The sentences should give a definition or state a fact; the questions should have mathematical significance, and preferably possess a certain degree of generality. [∉][4]

Exercise 2.2 Define five interesting finite sets. [∉]

---

[4] The symbol [∉] indicates that the exercise must be completed without using any mathematical symbol.

2.1 Sets

Exercise 2.3 The following expressions define sets. Turn words into symbols, using standard or Zermelo definitions. (Represent geometrical objects, e.g., planar curves, by their cartesian equations.)

1. *The set of negative odd integers.*
2. *The set of natural numbers with three decimal digits.*
3. *The set of rational numbers which are the ratio of consecutive integers.*
4. *The set of rational points in the closed unit cube.*
5. *The complement of the open unit disc in the complex plane.*
6. *The set of vectors of unit length in three-dimensional euclidean space.*
7. *The set of circles in the plane, passing through the origin.*
8. *The set of hyperbolae in the plane, whose asymptotes are the coordinate axes.*
9. *The set of lines tangent to the unit circle.*

Exercise 2.4 The following expressions define sets. Turn symbols into words. [↯]

1. $\{x \in \mathbb{Q} : 0 < x < 1\}$
2. $\{1/(2n+1) : n \in \mathbb{Z}\}$
3. $\{m2^{-k} : m \in 1 + 2\mathbb{Z}, \ k \in \mathbb{N}\}$
4. $\{x \in \mathbb{R} \setminus \mathbb{Z} : x^2 \in \mathbb{Z}\}$
5. $\{z \in \mathbb{C} \setminus \mathbb{R} : z^2 \in \mathbb{R}\}$
6. $\{z \in \mathbb{C} : |\text{Re}(z)| + |\text{Im}(z)| \leqslant 1\}$
7. $\{(m, n) \in \mathbb{Z}^2 : m \,|\, n\}$
8. $\{(x, y, z) \in \mathbb{R}^3 : x\,y\,z = 0\}$
9. $\{(x_1, \ldots, x_n) \in \mathbb{R}^n : \sum_k x_k = 0\}$
10. $\{x \in \mathbb{R} : \sin(2\pi x) = 0\}$
11. $\{(x, y) \in \mathbb{R}^2 : \sin(\pi x) \sin(\pi y) = 0\}$.

2.2 Functions

Functions are everywhere. Whenever a process transforms a mathematical object into another object, there is a function in the background. 'Function' is arguably the most used word in mathematics.

A **function** consists of two sets together with a rule[5] that assigns to *each* element of the first set a *unique* element of the second set. The first set is called the **domain** of the function and the second set is called the **co-domain**. A function whose domain is a set A may be called a function **over** A or a function **defined on** A. The terms **map** or **mapping** are synonymous with function. The term **operator** is used to describe certain types of functions (see below).

A function is usually denoted by a single letter or symbol, such as f. If x is an element of the domain of a function f, then the **value of** f **at** x, denoted by $f(x)$

---
[5] Below, we'll replace the term 'rule' with something more rigorous.

is the unique element of the co-domain that the rule defining f assigns to x. The notation

$$f : A \to B \qquad x \mapsto f(x) \qquad (2.16)$$

indicates that f is a function with domain A and co-domain B that **maps** $x \in A$ **to** $f(x) \in B$. The symbol x is the **variable** or (the **argument**) of the function. The symbols $\to$ and $\mapsto$ have different meanings, and should not be confused. The function

$$I_A : A \to A \qquad x \mapsto x$$

is called the **identity** (**function**) on A. When explicit reference to the set A is unnecessary, the identity is also denoted by Id or $\mathbb{1}$.

In Definition (2.16) the symbols used for the function's name and variable are inessential; the two expressions

$$f : \mathbb{R}\setminus\{0\} \to \mathbb{R} \quad x \mapsto \frac{1}{x} \qquad x : \mathbb{R}\setminus\{0\} \to \mathbb{R} \quad f \mapsto \frac{1}{f}$$

define exactly the same function (even though the rightmost expression breaks just about every rule concerning mathematical notation—see Sect. 6.2).

Let us use the word 'function' in short expressions. These are function definitions:

1. *The integer function that squares its argument.*
2. *The function that returns 1 if its argument is rational, and 0 otherwise.*
3. *The function that counts the number of primes smaller than a given real number.*
4. *The function that gives the distance between two points on the unit circle, measured along the circumference.*

We surmise that the function in item 2 is defined over the real numbers. Item 3 is a much-studied function in number theory. The set of values assumed by the function in item 4 is the closed interval $[0, \pi]$.

Functions of several variables are defined over cartesian products of sets. For example, the function

$$f : \mathbb{Z} \times \mathbb{Z} \to \mathbb{N} \qquad (x, y) \mapsto \gcd(x, y)$$

depends on two integer arguments, and hence is defined over the cartesian product of two copies of the integers. This definition requires a value for $\gcd(0, 0)$, which normally is taken to be zero.

Let $f : A \to B$ be a function. The set

$$\{(x, f(x)) \in A \times B : x \in A\} \qquad (2.17)$$

is called the **graph** of f. A function is completely specified by three sets: domain, co-domain and graph. We can now reformulate the definition of a function, replacing the imprecise term 'rule' with the precise term 'graph'. We write a formal definition.

2.2 Functions

DEFINITION. A **function** f *is a triple* (X, Y, G) *of non-empty sets. The sets X and Y are arbitrary, while G is a subset of $X \times Y$ with the property that for every $x \in X$ there is a unique pair $(x, y) \in G$. The quantity y is called the* **value of the function at** x, *denoted by* $f(x)$.

We see that, besides sets, the definition of a function requires the constructs of ordered pair and triple. It turns out that these quantities can be defined solely in terms of sets (see Exercise 2.14). So, to define functions, all we need are sets after all.

Given a function $f : A \to B$, and a subset $X \subset A$, the set

$$f(X) \stackrel{\text{def}}{=} \{f(x) : x \in X\} \tag{2.18}$$

is called the **image of X under** f. The assignment operator gives meaning to the symbolic expression $f(X)$, which otherwise would be meaningless, since we stipulated that the argument of a function is an element of the domain, not a subset of it. Thus $\sin(\mathbb{R})$ is the closed interval $[-1, 1]$.

Clearly, $f(X) \subset B$, and $f(A)$ is the smallest set that can serve as co-domain for f. The set $f(A)$ is often called the **image** or the **range** of the function f. This term is sometimes used to mean co-domain, which should be avoided. A **constant** is a function whose image consists of a single point.

The notation (2.18) is suggestive and widely used. However, in computer algebra, the quantities $f(x)$ and $f(X)$ (with x an element and X a subset of the domain, respectively) are written with a different syntax, e.g., `f(x)` and `map(f,X)` with Maple.

A function is said to be **injective** (or **one-to-one**) if distinct points of the domain map to distinct points of the co-domain. A function is **surjective** (or **onto**) if $f(A) = B$, that is, if the image coincides with the co-domain. A function that is both injective and surjective is said to be **bijective**.

For any non-empty subset X of the domain A, we define the **restriction of** f **to** X as

$$f|_X : X \to B \qquad x \mapsto f(x).$$

Given two functions $f : A \to B$ and $g : B \to C$, their **composition** is the function

$$g \circ f : A \to C \qquad x \mapsto g(f(x)). \tag{2.19}$$

The notation $g \circ f$ reminds us that f acts before g. The image $g(f(x))$ of x under $g \circ f$ is denoted by $(g \circ f)(x)$, where the parentheses isolate $g \circ f$ as the function's symbolic name. The hybrid notation $g \circ f(x)$ should be avoided.

If $f : A \to B$ is a bijective function, then the **inverse** of f is the function $f^{-1} : B \to A$ such that

$$f^{-1} \circ f = \mathbb{1}_A \qquad f \circ f^{-1} = \mathbb{1}_B$$

where $\mathbb{1}_{A,B}$ are the identities in the respective sets. A function is said to be **invertible** if its inverse exists. If $f : A \to B$ is injective, then we can always define the inverse of f by restricting its domain to $f(A)$ if necessary. In absence of injectivity, it may still be possible to construct an inverse by a suitable restriction of the function. Thus the arcsine may be defined by restricting the sine to the interval $[-\pi/2, \pi/2]$.

Let $f : A \to B$ be a function, and let C be a subset of B. The set of points

$$f^{-1}(C) \stackrel{\text{def}}{=} \{x \in A : f(x) \in C\} \tag{2.20}$$

is called the **inverse image** of the set C.

Since the definition of inverse image does not involve the inverse function, the inverse image exists even if the inverse function does not. These two concepts must be distinguished carefully. When the reciprocal of a function comes into play, things get very confusing, since we now have three unrelated objects represented by closely related notation:

$$f^{-1}(x) \qquad f^{-1}(\{x\}) \qquad f(x)^{-1}.$$

The first expression is well-defined if x belongs to the image of f and f is invertible there. In the second expression there is no condition on f, and x need only be an element of the co-domain. In the third expression the point x must belong to the domain of f, and $f(x)$ must be non-zero. Thus

$$\sin^{-1}(1) = \frac{\pi}{2} \qquad \sin^{-1}(\{1\}) = \frac{\pi}{2} + 2\pi\mathbb{Z} \qquad \sin(1)^{-1} = \csc(1).$$

In the first expression we tacitly assume that $\sin^{-1} = \arcsin : [-1, 1] \to [-\pi/2, \pi/2]$. In the third expression the symbol csc denotes the co-secant ($\csc(x) = 1/\sin(x)$), defined in the domain $\mathbb{R} \setminus \pi\mathbb{Z}$.

With a judicious use of definite and indefinite articles, we can specify a function's type without committing ourselves to a specific object.

1. The inverse of a trigonometric function.
2. The composition of a function with itself.
3. An integer-valued bijective function.
4. A function which coincides with its own inverse.

In item 2, we infer that the function maps its domain into itself. Functions of type 3 will be considered in the next section to define cardinality of sets. Functions of type 4 are called **involutions** (e.g., $x \mapsto -x$, over a suitable domain).

Writing about **real functions** is considered in Chap. 5.

2.2 Functions

Exercise 2.5 Turn symbols into words. [≢]

1. f^{-1}
2. $f^{-1}(0)$
3. $f(x^{-1})$
4. $f(0)^{-1}$
5. $f \circ f$
6. $(f \circ g)^{-1}$
7. $f|_{\mathbb{Z}}$
8. $f(\mathbb{R}^+)$
9. $f(\mathbb{R} \setminus \mathbb{Q})$
10. $f(A) \cap f(B)$
11. $f(\mathbb{R}) \cap \mathbb{Q}$
12. $\mathbb{Z} \cap f^{-1}(\mathbb{Z})$.

Exercise 2.6 Explain clearly and plainly. [≢, 50][6]

1. How do I divide two fractions?
2. I have a positive integer. How do I check if it's prime?
3. I have a positive integer. How do I check if it's a cube?
4. I have a positive integer. How do I check if it's the sum of two squares?
5. I have a cartesian equation of a circle, and a point. How do I check if the point lies inside the circle?
6. I have two lines in three-dimensional space. How do I check if they intersect?
7. I have two vectors on the plane. How do I check if they are linearly independent?
8. I have a list of quadratic polynomial functions and I must select the functions that assume both positive and negative values. What shall I do?
9. I have two real functions. From a sketch, it seems that their graphs intersect, and do so at a right angle. How do I verify that this is indeed the case?
10. I have three points on the plane. How do I compute the centre of the circle passing through them?
11. I have four points on the plane. How do I check if the points are vertices of a square?

Exercise 2.7 I have two finite sets and a function between them. I am able to compute the value of the function at each point of the domain, and to count and compare the elements of these sets. I need explicit instructions for answering the following questions. [≢, 50]

1. How do I check that my function is surjective?
2. How do I check that my function is injective?

Exercise 2.8 Answer each question as clearly as you can.

1. Let A and B be sets. Why are the sets $A^2 \setminus B^2$ and $(A \setminus B)^2$ not necessarily equal? Under what conditions are they equal?
2. Let $f : X \to Y$ be a function, and let A be a subset of X. Why are the sets A and $f^{-1}(f(A))$ not necessarily equal? Under what conditions are they equal?
3. Let A and B be subsets of the domain of a function f. Why are $f(A) \cap f(B)$ and $f(A \cap B)$ not necessarily equal? Under what conditions are they equal?

---

[6] Each assignment should contain no mathematical symbols and approximately 50 words.

2.3 Some Advanced Terms

We develop some advanced terminology and notation on sets. The content of this section is not essential for the rest of the book.

2.3.1 Families of Sets

The **power set** $\mathbf{P}(A)$ of a set A is the set of all subsets of A. Thus if $A = \{1, 2, 3\}$, then
$$\mathbf{P}(A) = \{\{\}, \{1\}, \{2\}, \{3\}, \{1, 2\}, \{1, 3\}, \{2, 3\}, \{1, 2, 3\}\}.$$

If A has n elements, then $\mathbf{P}(A)$ has 2^n elements. Indeed to construct a subset of A we consider each element of A and we decide whether to include it or leave it out, giving n binary choices. (A rigorous proof requires induction, see Exercise 8.4.)

A **partition** of a set A is a collection of pairwise disjoint non-empty subsets of A whose union is A. So a partition of A is a subset of $\mathbf{P}(A)$. For instance, the set $\{\{2\}, \{1, 3\}\}$ is a partition of $\{1, 2, 3\}$, and the even and odd integers form a partition of $\mathbb{Z}$. A partition may be described as a decomposition of a set into **classes**.

We write some phrases using these terms. Consider carefully the distinction between definite and indefinite articles (see Sect. 2.1.4 and Exercise 2.9).

The power set of a finite set.
The power set of a power set.
A partition of a power set.
The set of all partitions of a set.
A set of partitions of the natural numbers.
A finite partition of an infinite set.

Now some sentences:

Let us partition our interval into finitely many equal sub-intervals.
The plane may be partitioned into concentric annuli.
There is no finite partition of a triangle into squares.

2.3.2 Sums and Products of Sets

Let X and Y be sets of numbers. The **algebraic sum** $X + Y$ and **product** XY (also known as **Minkowski**[7] **sum** (**product**)), are defined as follows:

$$X + Y \stackrel{\text{def}}{=} \{x + y : x \in X, y \in Y\} \qquad XY \stackrel{\text{def}}{=} \{xy : x \in X, y \in Y\} \qquad (2.21)$$

---

[7] Hermann Minkowski (Polish: 1864–1909).

2.3 Some Advanced Terms

with the stipulation that repeated elements are to be ignored. For example, if $X = \{1, 3\}$ and $Y = \{2, 4\}$, then

$$X + Y = \{3, 5, 7\} \qquad XY = \{2, 4, 6, 12\}.$$

The expression 'sum of sets' is always understood as an algebraic sum. In the case of product, it is advisable to use the full expression to avoid confusion with the cartesian product.

If $X = \{x\}$ consists of a single element, then we use the shorthand notation $x + Y$ and xY in place of $\{x\} + Y$ and $\{x\}Y$, respectively (as we did in Sect. 2.1.2 for integers). For example

$$\frac{1}{2} + \mathbb{N} = \left\{\frac{1}{2}, \frac{3}{2}, \frac{5}{2}, \ldots\right\} \qquad \pi \mathbb{Z} = \{\ldots, -2\pi, \pi, 0, \pi, 2\pi, \ldots\}.$$

This notation leads to concise statements such as

$$m\mathbb{Z} + n\mathbb{Z} = \gcd(m, n)\mathbb{Z}$$

which combines algebraic sum and product of sets (see Exercise 7.5).

Elementary—but significant—applications of this construct are found in **modular arithmetic**. Let m be a positive integer. We say that two integers x and y are **congruent modulo** m if m divides $x - y$. This relation is denoted by[8]

$$x \equiv y \pmod{m}.$$

Thus

$$-3 \equiv 7 \pmod 5 \qquad 1 \not\equiv 12 \pmod 7.$$

The integer m is called the **modulus**. The set of integers congruent to a given integer is called a **congruence** (or **residue**) **class**. One verifies that the congruence class of k modulo m is the infinite set $k + m\mathbb{Z}$ given explicitly in (2.11). The congruence class of k modulo m is also denoted by $[k]_m$, $k \pmod m$, or, if the modulus is understood, by $[k]$ or $\bar{k}$.

The set of congruence classes modulo m is denoted by $\mathbb{Z}/m\mathbb{Z}$. If $m = p$ is a prime number, the notation $\mathbb{F}_p$ (meaning 'the field with p elements') may be used in place of $\mathbb{Z}/p\mathbb{Z}$. The set $\mathbb{Z}/m\mathbb{Z}$ contains m elements, which form a **partition** of $\mathbb{Z}$:

$$\mathbb{Z}/m\mathbb{Z} = \{m\mathbb{Z}, 1 + m\mathbb{Z}, 2 + m\mathbb{Z}, \ldots, (m-1) + m\mathbb{Z}\}.$$

Variants of this notation are used extensively in algebra, where one defines the sum/product of more general sets, such as groups and rings.

---

[8] This notation is due to Carl Friedrich Gauss (German: 1777–1855).

2.3.3 Representations of Sets

Consider the following geometrical sets:

The set of triangles with a vertex at the origin.
The set of triples of mutually tangent circles.

These definitions are easy to grasp, but how are we meant to work with sets of this kind? Suppose that we require a data structure suitable for computer implementation. We must then identify each element of our set with one or more concrete objects, such as numbers or matrices. This identification gives a description of a set in terms of another set, hopefully easier to handle.

Two sets A and B are said to be **equivalent** (written $A \sim B$) if there is a **bi-unique correspondence** between the elements of A and the elements of B, namely, if there exists a bijective function $f : A \to B$. A set equivalent to $\{1, 2, \ldots n\}$ is said to have **cardinality** n, and a set equivalent to $\mathbb{N}$ is said to be **countable** or **countably infinite**. The set $\mathbb{Z}$ is countable, and so is $m\mathbb{Z}$, for any $m \in \mathbb{N}$. A set X is **uncountable** if it contains a countably infinite subset Y, but X is not equivalent to Y. The set $\mathbb{R}$ is uncountable. We see that characterising the cardinality of infinite sets requires a more sophisticated approach than mere 'counting'.

A **representation** of a set A is any set B which is equivalent to A. (This is the most general acceptation of the term representation; in algebra, representations are based on a more specialised form of equivalence.)

For instance, the open unit interval and the real line are equivalent, as established by the bijective function

$$f : \mathbb{R} \to (0, 1) \qquad x \mapsto \frac{1}{\pi} \arctan(x) + \frac{1}{2}. \qquad (2.22)$$

Likewise, the exponential function establishes the equivalence $\mathbb{R} \sim \mathbb{R}^+$.

Let us consider representations of the set L of lines in the plane passing through a given point (a, b). The set L is uncountable. An element λ of L is an infinite subset of $\mathbb{R}^2$, which we write symbolically as

$$\lambda = \left\{ (x, y) \in \mathbb{R}^2 : y = b + s(x - a) \right\}$$

where s is a real number representing the line's slope. The line $x = a$ is not of this form, and must be treated separately. Collecting all the lines together, we obtain a symbolic description of L:

$$L = \left\{ \left\{ (x, y) \in \mathbb{R}^2 : y = b + s(x - a) \right\} : s \in R \right\} \cup \left\{ \left\{ (a, y) \in \mathbb{R}^2 : y \in \mathbb{R} \right\} \right\}.$$

2.3 Some Advanced Terms

The simple verbal definition of L seems to have drowned in a sea of symbols!

We look for a set equivalent to L with a more legible structure. An obvious simplification results from representing L as a set of **cartesian equations**:

$$L \sim \{y = b + s(x - a) : s \in \mathbb{R}\} \cup \{x = a\}.$$

We have merely replaced the solution set of an equation with the equation itself (cf. Sect. 3.3). This identification provides the desired bi-unique correspondence between the two sets.

We can simplify further. Because a and b are fixed, there is no need to specify them explicitly; it suffices to give the (possibly infinite) value of the slope. Alternatively, we could identify a line by an angle θ between 0 and π, measured with respect to some reference axis passing through the point (a, b). Because the angles 0 and π correspond to the same line, only one of them is to be included, resulting in the half-open interval $[0, \pi)$. The equivalence between $\mathbb{R} \cup \{\infty\}$ and $[0, \pi)$ may be achieved with a transformation of the type (2.22), where the included end-point 0 corresponds to the point at infinity.

Finally, any half-open interval may be identified with the **circle** $\mathbb{S}^1$, by gluing together the end-points of the interval. In our case this is achieved with the function $\theta \mapsto (\cos(2\theta), \sin(2\theta))$. The essence of our set is now clear:

$$L \sim \mathbb{R} \cup \{\infty\} \sim [0, \pi) \sim \mathbb{S}^1.$$

Exercise 2.9 Consider the phrases displayed in Sect. 2.3.1. Provide an example of each object being defined.

Exercise 2.10 Why are the sets $\emptyset$ and $\{\emptyset\}$ distinct? What are the elements of the power set $\mathbf{P}(\mathbf{P}(\mathbf{P}(\emptyset)))$? Explain.

Exercise 2.11 Represent the algebraic sum of sets as a function.

Exercise 2.12 Consider the function that performs the prime factorization of a natural number greater than 1. What would you choose for co-domain? Explain, discussing possible representations.

Exercise 2.13 Represent the following sets:

1. the set of open segments in the plane; the subset of segments of unit length;
2. the set of triangles with one vertex at the origin;
3. the polynomial set $\mathbb{Z}[x]$, as a set of integer sequences;
4. the real numbers, as a set of integer sequences;
5. the set of all finite subsets of $\mathbb{N}$, as a set of rational numbers.

Exercise 2.14 Prove that the definition

$$(a, b) \stackrel{\text{def}}{=} \{\{a\}, \{a, b\}\}$$

satisfies (2.4). (This shows that an ordered pair can be defined in terms of a set, so there's no need to introduce a new object.) Hence define an ordered triple in terms of sets.

Chapter 3
Essential Dictionary II

We further develop our mathematical dictionary, introducing the words **sequence**, **sum**, and **equation**. This will enable us to describe mathematical expressions of respectable complexity. As in the previous chapter, the last section contains advanced terms which are not essential for the rest of this book.

3.1 Sequences

A **sequence** is an ordered list of objects, not necessarily distinct, called the **terms** (or the **elements**) of the sequence. The terms of a sequence are represented by a common symbol, and each term is identified by an integer **subscript**:

$$(a_1, a_2, \ldots, a_n) \qquad (a_1, a_2, \ldots). \tag{3.1}$$

Here the common symbol is a, and the integer values assumed by the subscript begin from 1. The quantity a_1 reads "a sub 1", etc. Subscripts may also begin from 0, or from anywhere. The expression on the left denotes a finite sequence, the one on the right suggests that the sequence is infinite.

The **length** of a sequence is the number of its elements. Two sequences are equal if they have the same length, and if the corresponding terms are equal. If k is an unspecified integer, then a_k is called the **general term** of the sequence.

For example, the sequence of primes

$$(p_1, p_2, p_3, \ldots) = (2, 3, 5, \ldots)$$

is infinite. The general term p_k is the kth prime number.

There are other notations for sequences, displaying the general term alongside information about the subscript range:

$$(a_k)_{k=1}^n \qquad (a_k)_1^n \qquad (a_k)_{k=1}^\infty \qquad (a_k)_{k \geqslant 1} \qquad (a_k). \qquad (3.2)$$

There are also the **doubly-infinite sequences,** where the subscript runs through all the integers:

$$(a_k)_{k=-\infty}^\infty = (\ldots, a_{-1}, a_0, a_1, \ldots).$$

In Sect. 6.2 we shall discuss the usage of these notations.

The ellipsis is ubiquitous in sequence notation. When we use it, we must make sure that the missing terms are defined unambiguously. Thus the general term of the sequence of monomials

$$(2x, 2x^2, 2x^3, \ldots)$$

is clearly equal to $2x^k$. However, the expression

$$(3, 5, 7, \ldots)$$

is ambiguous, because there are several plausible alternatives for the identity of the omitted terms, such as $(9, 11, 13, \ldots)$ or $(11, 13, 17, \ldots)$. In the former case, we resolve the ambiguity by displaying the general term:

$$(3, 5, \ldots, 2k+1, \ldots).$$

In the latter, we need an accompanying sentence.

A **sub-sequence** of a sequence (a_k) is any sequence obtained from (a_k) by deleting terms. For instance, the primes that give remainder 1 upon division by 4 form a sub-sequence of the sequence of primes:

$$(5, 13, 17, 29, \ldots).$$

Some types of finite sequences have a dedicated terminology. A two-element sequence is an (**ordered**) **pair**, and a three-element sequence a **triple.** Occasionally one sees the terms **quadruple** or **quintuple** (I wouldn't go much beyond that), while an n-element sequence may be called an n-**uple**. A finite sequence of numbers may be called a **vector**, in which case we speak of **dimension** rather than length.

An infinite sequence $(a_1, a_2, \ldots)$ represents a **function** defined over the natural numbers. If the elements of the sequence belong to a set A, then such a function is defined as

$$a : \mathbb{N} \to A \qquad k \mapsto a_k.$$

We see that in the expression a_k, the symbol a is the function's name, the subscript k is an element of the domain, and $a_k \in A$ is the value $a(k)$ of the function at k. This

3.1 Sequences

interpretation clarifies the meaning of expressions such as a_{k^2}: it is the composition of two functions, much like $\sin(x^2)$.

Many familiar constructs involve sequences. Let A be a set of numbers, and let $(a_0, \ldots, a_n)$ be a finite sequence of elements of A with $a_n \neq 0$. A **polynomial** over A in the **indeterminate** x is an expression of the type

$$a_0 + a_1 x + a_2 x^2 + \cdots + a_n x^n. \tag{3.3}$$

In this notation, repeated addition is represented by the raised ellipsis. The elements of the sequence are called the **coefficients** of the polynomial, and the integer n is its **degree**. The coefficients a_0 and a_n are called, respectively, the **constant** and the **leading coefficient**. Each addendum in a polynomial is called a **monomial**, and a polynomial with two terms is a **binomial**. A polynomial of degree two is said to be **quadratic**; then we have **cubic**, **quartic**, and **quintic** polynomials. The set of all polynomials over the set A with indeterminate x is denoted by $A[x]$. For example

$$x^2 - x - 1 \in \mathbb{Z}[x] \qquad \frac{1}{2} - y^3 \in \mathbb{Q}[y].$$

A **rational function** is the ratio of two polynomials:

$$\frac{a_0 + a_1 x + \cdots + a_n x^n}{b_0 + b_1 x + \cdots + b_m x^m}. \tag{3.4}$$

Its degree is the largest of m and n (assuming that $a_n b_m \neq 0$). The set of all rational functions with coefficients in a set A and indeterminate x is denoted by $A(x)$.

A **multivariate polynomial** is a polynomial in more than one indeterminate. (The term **univariate** is used to differentiate from multivariate.)

$$x^2 y^2 - \frac{1}{2} x^4 - x y^3 \in \mathbb{Q}[x, y].$$

The **total degree** of each monomial is the sum of the degrees of the indeterminates, and the degree of a polynomial is the largest total degree among the monomials with non-zero coefficient. A multivariate polynomial is **homogeneous** if all monomials have the same total degree. The expression above may be described as

A homogeneous quartic polynomial in two indeterminates with rational coefficients.

EXAMPLE. Explain what is a polynomial. [✄]

A polynomial is a finite sum. Each term, called a monomial, is the product of a coefficient (typically, a real or complex number) and one or more indeterminates, each raised to some non-negative integer power.

3.2 Sums

Sums can be found in every corner of mathematics. We develop the dictionary and the very rich notation associated to summation.

Given a finite sequence of numbers $(a_1, \ldots, a_n)$, we form the sum of its elements:

$$\sum_{k=1}^{n} a_k = a_1 + a_2 + \cdots + a_n. \tag{3.5}$$

This notation was introduced by Fourier.[1] It is called the (delimited) **sigma-notation**, as it makes use of the capital Greek letter with that name, called the **summation symbol.** The subscript k is the **index** of summation, while 1 and n are, respectively, the **lower limit** and **upper limit** of summation. The quantity a_k is the **general term** of the sum. The integer sequence $(1, 2, \ldots, n)$, specifying the values assumed by the index of summation, is called the **range** of summation.

The summation index is a **dummy variable**. This a variable used for internal book-keeping (like an integration variable), and its identity does not affect the value of the sum:

$$\sum_{k=1}^{n} k^2 = \sum_{j=1}^{n} j^2 = 1^2 + 2^2 + \cdots + n^2.$$

I suggest that you adopt a summation symbol (one of i, j, k, l, m, n, see Sect. 6.2) and stick to it, unless there is a good reason to change it.

There are variants to the delimited sigma-notation (3.5). In an unrestricted sum, range information may be omitted altogether:

$$\sum_{k} \binom{n}{k} = 2^n \quad n \geq 0.$$

(This sum has in fact only finitely many non-zero terms.) The summation range may also be specified by inequalities placed below the summation symbols:

$$\sum_{1 \leq k \leq n} a_k \qquad \sum_{k \geq 1} a_k \qquad \sum_{1 \leq j, k \leq N} a_{j,k}. \tag{3.6}$$

Further conditions may be added to alter the range of summation:

$$\sum_{0 < |k| \leq 2} a_k = a_{-2} + a_{-1} + a_1 + a_2$$

---

[1] Jean Baptiste Joseph Fourier (French: 1768–1830).

3.2 Sums

$$\sum_{\substack{1 \leqslant k \leqslant 12 \\ \gcd(k,12)=1}} a_k = a_1 + a_5 + a_7 + a_{11}$$

$$\sum_{\substack{1 \leqslant k \leqslant 10 \\ k \text{ prime}}} a_k = a_2 + a_3 + a_5 + a_7.$$

The advantage of this notation—called the **standard form** of the sigma-notation—is that the summation index is no longer restricted to a sequence of consecutive integers.

EXAMPLE. Consider the following manipulation:

$$\sum_{-2 \leqslant k \leqslant n-3} 2^{k+2} = \sum_{0 \leqslant k+2 \leqslant n-1} 2^{k+2} = \sum_{0 \leqslant j \leqslant n-1} 2^j = 2^n - 1. \quad (3.7)$$

After adding 2 to each term in the inequalities, we have simply replaced $k+2$ with j, obtaining the sum of a geometric progression which is evaluated explicitly. With this notation the change of summation index is unproblematic.

EXAMPLE. For any natural number n, let $\varphi(n)$ be the number of positive integers smaller than n and relatively prime to it (with $\varphi(1) = 1$). Thus $\varphi(12) = 4$. This is **Euler's φ-function**[2] of number theory, which is defined in symbols as follows:

$$\varphi(1) = 1 \qquad \varphi(n) = \sum_{\substack{1 \leqslant k < n \\ \gcd(k,n)=1}} 1, \qquad n > 1. \quad (3.8)$$

Simply by letting $a_k = 1$ in (3.5), we have turned a summation into an algorithm that counts the elements of the set specified by the given conditions. This neat device illustrates the flexibility of the sigma-notation. Alternatively, Euler's function may be defined as the cardinality of a set, using the cardinality symbol '#' (see Sect. 2.1):

$$\varphi(1) = 1 \qquad \varphi(n) = \#\{k \in \mathbb{N} : k < n, \ \gcd(k,n) = 1\} \qquad n > 1.$$

Sums may be nested. A **double sum** is defined as follows:

$$\sum_{j=1}^{J} \sum_{k=1}^{K} a_{j,k} \stackrel{\text{def}}{=} \sum_{j=1}^{J} \left(\sum_{k=1}^{K} a_{j,k} \right)$$

$$= \sum_{k=1}^{K} a_{1,k} + \sum_{k=1}^{K} a_{2,k} + \cdots + \sum_{k=1}^{K} a_{J,k}.$$

---

[2] Leonhard Euler (Swiss: 1707–1783). An accessible account of Euler's mathematics is found in [11].

The sum in parentheses is a function of the outer summation index j; this sum is performed repeatedly, each time with a different value of j. The use of matching symbols for the index and the upper limit of summation (j, J, k, K) is particularly appropriate here. If the two ranges of summations are independent, inner and outer sums can be swapped. The commutative and associative laws of addition ensure that the value of the sum will not change. We illustrate this process with an example.

$$\sum_{j=0}^{1} \sum_{k=1}^{3} a_{j,k} = (a_{0,1} + a_{0,2} + a_{0,3}) + (a_{1,1} + a_{1,2} + a_{1,3})$$
$$= (a_{0,1} + a_{1,1}) + (a_{0,2} + a_{1,2}) + (a_{0,3} + a_{1,3})$$
$$= \sum_{k=1}^{3} \sum_{j=0}^{1} a_{j,k}.$$

This chain of equalities adopts a standard layout and alignment. If the indices in a double sum have the same range, then they may be grouped together:

$$\sum_{i,j=1}^{N} a_{i,j} = \sum_{i=1}^{N} \sum_{j=1}^{N} a_{i,j}. \tag{3.9}$$

A sum with infinitely many summands is called a **series**:

$$\sum_{k=1}^{\infty} a_k = a_1 + a_2 + \cdots. \tag{3.10}$$

The summation range may be specified in several ways:

$$\sum_{k \geq 1} a_k \qquad \sum_{k=-\infty}^{\infty} a_k \qquad \sum_{k \in \mathbb{Z}} a_k \qquad \sum_{k} a_k.$$

We haven't yet explained what (3.10) means. We write the definition using the assignment operator:

$$\sum_{k=1}^{\infty} a_k \stackrel{\nabla}{=} \lim_{n \to \infty} \sum_{k=1}^{n} a_k. \tag{3.11}$$

The above limit—if it exists—is called the **sum of the series**, and the series is said to **converge**. Otherwise the series **diverges**. If a series has non-negative terms, then convergence is sometimes expressed with the suggestive notation (cf. (2.3), p. 11)

$$\sum_{k \geq 0} a_k < \infty. \tag{3.12}$$

3.2 Sums

A **power series** is a series whose terms feature increasing powers of an indeterminate:

$$\sum_{k \geq 0} a_k x^k. \tag{3.13}$$

For the values of x for which the series (3.13) converges, the power series represents a function.

The terminology and notation introduced for sums extends with obvious modifications to products:

$$\prod_{k=1}^{n} a_k = a_1 \cdot a_2 \cdots a_n. \tag{3.14}$$

An infinite product is called just that—there is no special name for it. Its value is defined as the limit of a sequence of finite products, as we did for sums (3.11). Infinite products are often written in the form

$$\prod_{k \geq 0} (1 + a_k)$$

because for convergence we must have $a_k \to 0$.

All constructs introduced above may be extended to combinations of sums and products.

Exercise 3.1 Write out the following sums in full.

1. $\sum\limits_{0 \leq k-1 < 3} a_k$
2. $\sum\limits_{k^2 < 9} a_{-k}$
3. $\sum\limits_{k^2 \leq k+2} a_{1-k}$
4. $\sum\limits_{\substack{k \in 2\mathbb{Z}+1 \\ |k|<5}} a_k$
5. $\sum\limits_{\substack{|k-3|<5 \\ \gcd(k,6)>1}} a_k$
6. $\sum\limits_{k^2 \leq 9} a_{k^2}$

3.3 Equations and Identities

Let f and g be functions with the same domain X and co-domain Y. An **equation** (**on** or **over** X) is an expression of the type

$$f(x) = g(x). \tag{3.15}$$

The quantity x is the equation's **unknown**.

$$\begin{array}{ll} x^2 - 2x - 4 = 0 & \text{An algebraic equation.} \\ \cos(x) = \sin(x) & \text{A trigonometric equation.} \\ \log(1+x) = -x & \text{A transcendental equation.} \end{array} \tag{3.16}$$

These are equations over (a subset of) $\mathbb{R}$, and each equation is described in broad terms by an attribute (algebraic, trigonometric, transcendental). We shall learn more about these attributes in Sect. 3.4.2.

The expression (3.15) defines a property that each point $x \in X$ either has or doesn't have.[3] This prompts the definition of the **solution set** of Eq. (3.15), given by

$$\{x \in X : f(x) = g(x)\}.$$

For instance, over the real numbers the solution set of the first equation in (3.16) is $\{1 + \sqrt{5}, 1 - \sqrt{5}\}$, while over $\mathbb{Q}$ the solution set is empty. We see that the solution set of an equation depends on the ambient set.

If the co-domain Y of f and g is a set of numbers (e.g., $Y = \mathbb{Z}, \mathbb{Q}, \mathbb{R}, \mathbb{C}$), then, by replacing f by $f - g$, we can reduce Eq. (3.15) to the simpler form

$$f(x) = 0. \tag{3.17}$$

An element x of the solution set of this equation is called a **zero** of f, but if $f(x)$ is a polynomial, we speak of a **root** of f. We also say that f **vanishes at** x. A function f **vanishes identically** on a set if it vanishes at every point of this set, in which case we use the emphatic notation $f(x) \equiv 0$. For example, the real function $x \mapsto \sin(\pi x)$ vanishes identically on $\mathbb{Z}$.

More generally, an equation may be reduced to the form (3.17) if the co-domain of f and g is a **group** with respect to addition. In this case the zero on the right-hand side is the zero element of the group. This is not necessarily the number 0, but it could be a zero matrix, a zero polynomial, a zero function, etc.

An example is given by the **differential equations**, which relate a function to its derivatives:

$$\frac{dx}{dt} - x = 0 \qquad t^2 \frac{d^2x}{dt^2} + t \frac{dx}{dt} + (t^2 - 1)x = 0. \tag{3.18}$$

Here the unknown is x (not t), where $x = x(t)$ is a **function**, and the underlying ambient set is a **set of differentiable functions**. The symbol 0 represents the **zero function**, namely the constant function that assumes the value zero everywhere. We'll consider other types of equations in Sect. 3.5.

An **identity** (or **indeterminate equation**) is an equation whose solution set is equal to the ambient set:

$$(x - 1)^3 = x^3 - 3x^2 + 3x - 1.$$

After simplification, every identity reduces to the standard form $0 = 0$. This doesn't mean that identities are trivial, far from it; identities express equivalence of functions. However, they are ephemeral quantities, which disappear if they are simplified.

---

[3] In Chap. 4 we shall see that an equation is a special type of **predicate**, which in turn is a special type of **function**.

3.3 Equations and Identities

EXAMPLE. The identity

$$x^{2^n} - y^{2^n} = (x-y)\prod_{k=0}^{n-1}\left(x^{2^k} + y^{2^k}\right)$$

gives the full factorisation of the difference of two monomials whose degree is a power of 2, into the product of polynomials with integer coefficients.

EXAMPLE. Over the set $\mathbb{R}^2$, the expression $x + y = y + x$ is an identity, representing the commutativity of the addition of real numbers. The similar expression $x + y = 1 - y$ is an equation, whose solution set is a line in $\mathbb{R}^2$.

By restricting the ambient set to the solution set, every equation becomes an identity. Clearly, $\sin(\pi x) = 0$ is an equation over $\mathbb{R}$ and an identity over $\mathbb{Z}$. For a more subtle example, consider the equation $f(x) = x^5 - x = 0$. The factorisation $x^5 - x = x(x-1)(x+1)(x^2+1)$ shows that the solution set over $\mathbb{C}$ is $\{0, \pm 1, \pm\sqrt{-1}\}$, a set with five elements. Consider now the set

$$X = \mathbb{Z}/5\mathbb{Z} = \{0 + 5\mathbb{Z},\ 1 + 5\mathbb{Z},\ 2 + 5\mathbb{Z},\ 3 + 5\mathbb{Z},\ 4 + 5\mathbb{Z}\}$$

of congruence classes modulo 5. Let us evaluate our function f at all points of $\mathbb{Z}/5\mathbb{Z}$, writing k for $k + 5\mathbb{Z}$:

$$f(0) = 0 \equiv 0 \pmod{5}$$
$$f(1) = 0 \equiv 0 \pmod{5}$$
$$f(2) = 30 \equiv 0 \pmod{5}$$
$$f(3) = 240 \equiv 0 \pmod{5}$$
$$f(4) = 1020 \equiv 0 \pmod{5}.$$

The function f vanishes identically over $\mathbb{Z}/5\mathbb{Z}$, and hence the expression $x^5 = x$ is an identity!

An expression of the type

$$\begin{cases} f_1(x) = 0 \\ f_2(x) = 0 \\ \vdots \\ f_n(x) = 0 \end{cases} \qquad x = (x_1, \ldots, x_m) \tag{3.19}$$

where all functions have the same domain and co-domain, is called a **system of n simultaneous equations in m unknowns**. The solution set of a system of equations is the intersection of the solution sets of the individual equations.

EXAMPLE. Explain what is an equation, and its solutions. [≢]

BAD: *An equation is when we equate two functions. The solution is when the functions are the same.*

The inappropriate use of 'when' is easily spotted (see Sect. 1.1), but there is a more serious flaw. The expression 'equating two functions' means that we seek conditions under which the two functions become the same function. That's not what we had in mind. The operands of the equal sign in expression (3.15) are not functions, but rather values of functions.

GOOD: *An equation is an expression that identifies the value of two functions at a generic point of their common domain. The solutions of an equation are the points at which the two functions assume the same value.*

(The expression 'equating two functions' may be appropriate for **functional equations**, see (3.21).)

3.4 Expressions

The generic term **expression** indicates the symbolic encoding of a mathematical object. For instance, the string of symbols '$2 + 3$' is a valid expression, and so is '$x \mapsto f(x)$', while '$2 + \times 3$' is incorrect and does not represent an object.

It would seem that any correct expression should have—in principle, at least—a unique **value**, representing some agreed 'fully simplified' form of the expression. For instance, it could be argued that the value of $13/91$ is $1/7$, and that of $\sqrt{2187}$ is $27\sqrt{3}$. The simplified value is more concise and informative, and would enable us—among other things—to recognise when two expressions represent the same object.

This is not so simple. For example, the two expressions

$$\sqrt{3} - \sqrt{2} \qquad \sqrt{5 - 2\sqrt{6}}$$

have the same value, yet there is no compelling reason for choosing one over the other. The same object may be viewed from different angles, and our choice of representation will depend on the context.

The following well-known identity makes an even stronger point:

$$\frac{1 - x^n}{1 - x} = 1 + x + x^2 + \cdots + x^{n-1}.$$

The right-hand side—the sum of n monomials—is the 'fully simplified' version, while the 'unsimplified' left-hand side has only four terms. While simplification reveals the true nature of the object—a polynomial as opposed to a rational function—it's the unsimplified value which provides information about the sum.

3.4 Expressions

Given that agreeing on a unique value of an expression proves difficult, we shift our attention to the **type** of value (set, number, function, etc.), which assigns the expression to a certain class of objects. This class provides the broadest characterisation of the object in question. Again, we must exercise some judgement. The expressions

$$1+1 \qquad \int_0^\pi \sin(x)\,\mathrm{d}x$$

have the same value, but their structure is so different that the coincidence of their values seems secondary. Whereas the expression on the left is unquestionably 'a number', or 'a positive integer', that on the right is 'a definite integral.' On the other hand, there may be circumstances in which the reductionist description of the integral as a number is appropriate, for instance when discussing integrability of functions.

Keeping in mind the difficulties mentioned above, we now turn to the description of mathematical expressions.

3.4.1 Levels of Description

We develop the idea of successive refinements in the description of an expression, from the general to the particular. The appropriate level of details to be included will vary, depending on the situation. We treat in parallel verbal and symbolic descriptions, as far as it is reasonable to do so.

Let us consider the definition of a set. The coarsest level of description is

$$\{\ldots\} \qquad \text{A set}$$

where the object's type is identified by the curly brackets. The use of the indeterminate article—'a' set rather than 'the' set—reflects our incomplete knowledge.

The next level in specialisation identifies the ambient set:

$$\{(x, y) \in \mathbb{Z}^2 : \ldots\} \qquad \text{A set of integer pairs}$$

Now we begin to build the defining properties of our set:

$$\{(x, y) \in \mathbb{Z}^2 : \gcd(x, y) = 1, \ldots\} \qquad \text{A set of pairs of co-prime integers}$$

The final step completes the definition:

$$\{(x, y) \in \mathbb{Z}^2 : \gcd(x, y) = 1, \ 2|xy\} \qquad \text{The set of pairs of co-prime integers, with exactly one even component}$$

Accordingly, the indefinite article has been replaced by the definite article. Now both words and symbols describe one and the same object, and one should consider the relative merits of the two presentations. The term 'exactly' is, strictly speaking, redundant, but it helps the reader realise that x and y cannot both be even. A robotic translation of symbols into words

> The set of elements of the cartesian product of the integers with themselves, whose components have greatest common divisor equal to 1, and such that 2 divides the product of the components

lacks the synthesis that comes with understanding.

Expressions may be **nested**, like boxes within boxes. We begin with the two expressions

$$(\cdots)^2 \qquad \text{A square}$$
$$\sum \cdots \qquad \text{A sum}$$

We only see the outer structure of these objects. We compose them in two different ways:

$$\left(\sum \cdots\right)^2 \qquad \text{The square of a sum}$$
$$\sum (\cdots)^2 \qquad \text{A sum of squares}$$

The first term in each expression identifies the object's outer layer. There is still one indefinite article, reflecting a degree of generality. We specialise further:

$$\sum_{n=1}^{\infty} \left(\frac{1}{n}\right)^2 \qquad \text{The sum of the square of the reciprocal of the natural numbers}$$

Words or symbols now define a unique object, with three levels of nesting. By contrast, in the nested expression

$$\left(\sum_{n=1}^{\infty} a_n\right)^2 \quad a_n \in \mathbb{Q} \qquad \text{The square of the sum of the elements of a rational sequence}$$

the innermost object—a rational sequence—is still generic.

In these examples words and symbols are interchangeable; in actual writing, some concepts are best expressed with words, others with symbols, while most of them require both. For instance, the symbolic expression

$$(1 - x, 2x + x^2, \ldots, nx^{n-1} + (-x)^n, \ldots) \qquad (3.20)$$

3.4 Expressions

defines an infinite sequence succinctly and unambiguously. Using words, we could begin to describe it as follows:

A sequence
An infinite sequence
An infinite sequence of binomials

Increasing further the accuracy of the verbal description is pointless, since the symbolic expression (3.20) is clearly superior in delivering exact information. On the other hand, words can place this expression *in a context,* which is something symbols cannot do. To illustrate this point, we supplement the description given above with additional information which emphasises a particular property.

An infinite sequence of binomials
  with integer coefficients
  with unbounded coefficients
  with increasing degree
  whose leading term alternates in sign.

3.4.2 Describing Expressions

We expand our dictionary with terms describing broad attributes of expressions.

An expression involving numbers, the four arithmetic operations, and raising to an integer or fractional power (extraction of roots), is called an **arithmetical expression.** The value of an arithmetical expression is a number. A combination of rational numbers and square roots of rational numbers is called a **quadratic irrational** or a **quadratic surd**. The following expressions are arithmetical.

$191861^2 - 3 \cdot 110771^2 = -2$ *An arithmetic identity, with a surprising cancellation.*

$\dfrac{3 + 2\sqrt{2}}{8 - 3\sqrt{7}}$ *The ratio of two quadratic surds having distinct radicands.*

If indeterminates are present, we speak of an **algebraic expression**.

$\dfrac{\sqrt[6]{ab - (ab)^{-1}}}{\sqrt[3]{a^2 b^2 + ab + 1}}$ *An algebraic expression with two indeterminates and higher-index roots.*

Polynomials and rational functions are algebraic expressions (see Sect. 3.1). They may also be characterised as **rational expressions**, since they don't involve fractional powers of the indeterminates.

$$\frac{1}{x+1} + \frac{1}{x^2+1} + \cdots + \frac{1}{x^n+1} \qquad \text{The sum of finitely many rational functions, with increasing degree.}$$

$$\frac{\left((x^2+1)^2+1\right)^2+1}{(x^2+1)^2+1} \qquad \text{A rational expression, involving repeated compositions of a polynomial function with itself.}$$

The following mathematical Russian doll

$$\sqrt{x + \sqrt{2x + \sqrt{3x + \cdots + \sqrt{nx}}}} \qquad n \in \mathbb{N}$$

could be described as

An algebraic expression in one indeterminate, featuring a finite number of nested square roots.

The functions sine, cosine, tangent, secant, etc., are called **trigonometric functions** (or **circular functions**). A **trigonometric expression** is an expression containing trigonometric functions.

$$8\cos(z)^4 - 8\cos(z)^2 + 1 \qquad \text{A quartic trigonometric polynomial.}$$

Trigonometric functions belong to the larger class of **transcendental functions**, which are functions not definable by an algebraic expression (the exponential, the logarithm, etc.). In the expressions (3.16), we have used the terms algebraic, trigonometric and transcendental to describe equations.

The term **analytical expression** is used in the presence of infinite processes.

$$\sqrt[3]{1+x} = 1 + \frac{1}{3}x - \frac{1}{9}x^2 + \frac{5}{81}x^3 + \cdots \qquad \text{The first few terms of the series expansion of an algebraic function.}$$

$$\lim_{n \to \infty} \left(1 + \frac{1}{n}\right)^n = e \qquad \text{Napier's constant as the limit of a rational sequence.}$$

$$2 \prod_{k=1}^{\infty} \frac{(2k)^2}{(2k)^2 - 1} = \pi \qquad \text{An infinite product formula for Archimedes' constant.}$$

An **integral expression** is an expression involving integrals.

3.4 Expressions

$$\ln(x) = \int_1^x \frac{1}{t}\,dt \qquad \text{An integral expression for the natural logarithm.}$$

The term **combinatorial** is appropriate for expressions involving counting functions, such as the **factorial** function, or the **binomial coefficient**.

$$\frac{1}{2^{2k}}\frac{(2k)!}{(k!)^2} \qquad \text{A rational combination of exponentials and factorials.}$$

$$\sum_{k=0}^{n}\binom{n+k}{k} \qquad \text{A finite sum of binomial coefficients.}$$

In Chap. 4 we deal with the expressions found in logic: the **boolean** expressions.

Exercise 3.2 For each expression, provide two levels of description: [✦]

(i) a coarse description, which only identifies the object's type (set, function, equation, etc);
(ii) a finer description, which defines the object in question, or characterises its structure.

1. $3^3 + 4^3 + 5^3 = 6^3$
2. $\sqrt{3} + \sqrt[3]{3} + \sqrt[4]{3}$
3. $\frac{7}{5} < \sqrt{2} < \frac{17}{12}$
4. $x^3 - x - 1$
5. $xy > 0$
6. $(x+y)^3 = x^3 + 3x^2y + 3xy^2 + y^3$
7. $y - x^2 - x = 0$
8. $1 = 0$
9. $\sin(x - y) = \sin(x)\cos(y) - \cos(x)\sin(y)$
10. $\frac{d^2x}{dt^2} - 2\frac{dx}{dt} - 3 = 0$
11. $f(x, y) \leqslant g(x, y)$
12. $\begin{cases} f(x, y) = 0 \\ g(x, y) = 0 \end{cases}$
13. $(A \cup B) \cap C = (A \cap C) \cup (B \cap C)$
14. $\frac{1}{1^4} + \frac{1}{2^4} + \frac{1}{3^4} + \cdots = \frac{\pi^4}{90}$.

Exercise 3.3 Same as in Exercise 3.2.

1. $2\mathbb{Z} \setminus 4\mathbb{Z}$
2. $\{\{\emptyset\}\}$
3. $(\mathbb{Q} \setminus \mathbb{Z})^2$
4. $\mathbb{R}^2 \setminus \mathbb{Q}^2$
5. $(\mathbb{R} \times \mathbb{Z}) \cup (\mathbb{Z} \times \mathbb{R})$

6. $1 + 2(1 + 2(1 + 2\mathbb{Z}))$
7. $(\{a\}, \{a\}, \{a\}, \ldots)$
8. $(1, -3, 5, \ldots, (-1)^n(2n+1), \ldots)$
9. $(1, 2^{2k}, 3^{2k}, \ldots, n^{2k})$
10. $\{(1, 3), (3, 5), (5, 7), \ldots\}$
11. $((x_1), (x_1, x_2), (x_1, x_2, x_3), \ldots)$
12. $(\{\omega\}, \{\{\omega\}\}, \{\{\{\omega\}\}\}, \ldots)$.

Exercise 3.4 Same as in Exercise 3.2.

1. $f : \mathbb{R} \to \mathbb{R}, \quad x \mapsto x + 1$

2. $\displaystyle\int_0^\infty \frac{f(x)}{g(x)} \mathrm{d}x$

3. $\displaystyle\frac{\mathrm{d}}{\mathrm{d}x} f(x)g(x) = \frac{\mathrm{d}f(x)}{\mathrm{d}x} g(x) + \frac{\mathrm{d}g(x)}{\mathrm{d}x} f(x)$

4. $\displaystyle f(x) = \int_0^x \frac{\mathrm{d}f(y)}{\mathrm{d}y} \mathrm{d}y$

5. $\displaystyle\int F(x, y) \mathrm{d}x$

6. $\displaystyle\int_0^1 \mathrm{d}x \int_0^1 H(x, y) \mathrm{d}y$

7. $\displaystyle\cos(x) = \sum_{k=0}^\infty \frac{(-1)^k x^{2k}}{(2k)!}$

8. $\displaystyle\frac{\partial F(x, y, z)}{\partial x} + \frac{\partial F(x, y, z)}{\partial y} + \frac{\partial F(x, y, z)}{\partial z}$

9. $\displaystyle\prod_{k=1}^n \frac{\partial \mathscr{L}(\theta_1, \ldots, \theta_n)}{\partial \theta_k}$

10. $\displaystyle\sum_{k \geqslant 1} x^{n^2}$

11. $\displaystyle\int_0^\infty x^{t-1} \mathrm{e}^{-x} \mathrm{d}x$

12. $\displaystyle\pi x \prod_{n \geqslant 1} \left(1 - \frac{x^2}{n^2}\right)$.

3.5 Some Advanced Terms

We introduce some advanced words and symbols concerning sets, sequences, and equations.

3.5.1 Sets and Sequences

Let (A_k) be a **sequence of sets**, which may be finite or infinite. The binary set operations of union and intersection generalise to an arbitrary number of operands:

$$\bigcup_k A_k = A_1 \cup A_2 \cup \cdots \qquad \bigcap_k A_k = A_1 \cap A_2 \cap \cdots .$$

These expressions denote the set of elements belonging to at least one of the sets A_k, and to every one of the sets A_k, respectively. The integer index conforms to the notational rules for sums and products (Sect. 3.2):

$$\bigcap_{k=2}^{\infty} \Theta_k \qquad \bigcup_{\substack{k \in 2\mathbb{N} \\ k \geq K}} (A_k \setminus B_k).$$

However, infinite unions and intersections may also be controlled by a *real* (as opposed to integer) index, which adds flexibility to this notation:

$$\Omega = \bigcup_{\alpha > 0} \Omega_\alpha \qquad \Omega_\alpha = \{(x, y) \in \mathbb{R}^2 : y = x^2 + \alpha x\}.$$

(Sketch the set Ω.)

A sequence of sets is **descending** (or **nested**) if

$$A_1 \supset A_2 \supset A_3 \supset \cdots$$

and **ascending** if

$$A_1 \subset A_2 \subset A_3 \subset \cdots .$$

EXAMPLE. The following symbols

$$m\mathbb{Z} \supset m^2\mathbb{Z} \supset \cdots \supset m^k\mathbb{Z} \supset \cdots \qquad \bigcap_{k \geq 1} m^k\mathbb{Z} = \begin{cases} \{0\} & \text{if } |m| \neq 1 \\ \mathbb{Z} & \text{if } |m| = 1 \end{cases}$$

tell a story. Say it with words. [∉]

The set of multiples of a power of an integer contains the set of multiples of any higher power of the same integer. By considering increasing powers we obtain an infinite nested sequence of sets. Apart from a trivial case, the only element common to all these sets is the origin.

EXAMPLE. The **Farey sequence**[4] ($\mathscr{F}_n$) is an ascending sequence of sets. Its general term $\mathscr{F}_n$ is the set of all reduced fractions in the unit interval whose denominator does not exceed n:

$$\mathscr{F}_1 = \{0, 1\}$$
$$\mathscr{F}_2 = \left\{0, \frac{1}{2}, 1\right\}$$
$$\mathscr{F}_3 = \left\{0, \frac{1}{3}, \frac{1}{2}, \frac{2}{3}, 1\right\}$$
$$\mathscr{F}_4 = \left\{0, \frac{1}{4}, \frac{1}{3}, \frac{1}{2}, \frac{2}{3}, \frac{3}{4}, 1\right\}$$
$$\vdots$$

Let us now consider the representation of **sets of sequences**. Let A be any set. We form the set of all *finite* sequences of elements of A, with n terms. This set is the cartesian product A^n of n copies of the set A: the first element a_1 of a sequence is chosen from the first copy of A, the second element from the second copy of A, and so on. For instance, the set $\{0, 1\}^n$ is the set of all binary sequences with n-digits, while $\mathbb{Q}^n$ is the set of all n-uples of rational numbers. It follows that the infinite union

$$\bigcup_{n \geqslant 1} \mathbb{Z}^n$$

represents the set of all finite integer sequences.

The set of all infinite sequences of elements of A has the structure of a cartesian product with infinitely many terms. It would seem natural to denote it by A^∞. However, the idiomatic notation $A^\mathbb{N}$ is more common, due to its greater flexibility. Thus $A^\mathbb{Z}$ denotes the set of doubly-infinite sequences of elements of A, and one even finds $A^{\mathbb{Z}^2}$ for the set of two-dimensional arrays: $(a_{i,j}) \in A^{\mathbb{Z}^2}$.

3.5.2 More on Equations

An equation whose unknown is a function is called a **functional equation**.

$$f(x) = f(x+1) \qquad f(f(x)) = x. \qquad (3.21)$$

---

[4] This sequence, named after John Farey, Sr. (English: 1766–1826), was actually discovered by the Frenchman Charles Haros.

3.5 Some Advanced Terms

Here the unknown is f, not x. To make this plain, let $\mathbb{1}$ be the identity function in the appropriate set of functions, and let τ be the function that increases its argument by 1: $\tau(x) = x + 1$. Equations (3.21) may now be rewritten without reference to the function's argument, using the composition operator—see (2.19).

$$f = f \circ \tau \qquad f \circ f = \mathbb{1}. \tag{3.22}$$

Solutions of these equations are **periodic functions** and **involutions**, respectively.

The class of functional equations includes the **differential equations** (which we've seen in Sect. 3.3) and the **integral equations**. The latter are equations in which the unknown function appears under an integral sign, e.g.,

$$\sqrt{\pi} f(x) = \int_0^\infty e^{-xy} f(y) dy.$$

An equation whose unknown is a set is called a **set equation**:

$$(X + X) \cap X = \emptyset. \tag{3.23}$$

Here the symbol X necessarily represents a set, and hence $X + X$ is an algebraic sum of sets [defined in (2.21)]. This set equation may be defined over any family of sets of numbers, such as $\mathbf{P}(\mathbb{Z})$, the power set of the integers. A solution of this equation is called a **sum-free set**; thus the set of odd integers is sum-free.

It should be clear that equations can be written having any type of mathematical object as unknown.

Exercise 3.5 Same as in Exercise 3.2.

1. $\bigcap_k f^{-1}(A_k)$
2. $|\bigcup_{j,k} P_{j,k}|$
3. $f\left(\bigcup_\alpha \Theta_\alpha\right) = \bigcup_\alpha f(\Theta_\alpha)$
4. $(\mathbb{Z} + \sqrt{2}\mathbb{Z})^n$
5. $\{1, x, x^2\}^{\mathbb{N}}$
6. $\mathbb{Z}^{\mathbb{Z}}$
7. $X = \{X\}$
8. $\Gamma(z + 1) = z\Gamma(z)$
9. $\underbrace{f \circ f \circ \cdots \circ f}_{n} = \mathbb{1}$
10. $\underbrace{[0, 1] + [0, 1] + \cdots + [0, 1]}_{n}$.

Exercise 3.6 Write an equation whose unknown is (i) a matrix; (ii) a polynomial; (iii) a sequence.

Exercise 3.7 Ask some questions about the Farey sequence.

Exercise 3.8 Let X, Y be sets, and let $\mathscr{F}, \mathscr{G} : X \to \mathbf{P}(Y)$. Prove that the set equation $\mathscr{F}(x) = \mathscr{G}(x)$ is equivalent to the equation

$$\mathscr{F}(x) \Delta \mathscr{G}(x) = \emptyset$$

in the sense that the two equations have the same solution set. Thus every set equation may be reduced to the form $\mathscr{F}(x) = \emptyset$, which resembles (3.17).

Chapter 4
Mathematical Sentences

Consider the following theorem of analysis:

$$\text{Let } f : \mathbb{R} \to \mathbb{R} \text{ be a differentiable function. Then } f \text{ is continuous.} \qquad (4.1)$$

This statement comprises two sentences. The first sentence does not state a fact: it's an assumption. We aren't told which function f is, so there is no question of this statement about f being true or false. But still we use it as a basis for the rest of the argument. The second sentence does just that: it's a deduction of a new fact from the assumption. The symbol f is an internal variable of the statement, and we can make it disappear:

Every differentiable real function is continuous.

This statement is equivalent to (4.1).

In this chapter we begin to analyse mathematical sentences, and for this purpose we need some elements of logic. The chief attribute of a mathematical statement is its truth or falsehood. Accordingly, we define the two **logical** (or **boolean**) **constants** TRUE and FALSE, abbreviated T, F or 1, 0, respectively. The simplest sentences are the **relational expressions**, formed by combining pairs of mathematical objects via **relational operators**. The value of a relational expression is a logical constant. To form more complex sentences we combine relational expressions using **logical operators**, much like combining numbers using addition and multiplication. The formulation of high-level statements such as (4.1) requires logical functions—called **predicates**—and **quantifiers**, which resemble integrals.

These constructs have accompanying symbols, and the symbols of logic are cryptic and alluring. While they help us understand the structure of sentences, they don't necessarily help us write good sentences. Excessive use of logical symbols clutters the exposition, obscuring meaning. A recurrent theme of this chapter is that, quite often, concepts of great logical depth are better expressed with words.

4.1 Relational Operators

We begin with a simple symbolic sentence:

$$0 < 1. \tag{4.2}$$

This is a **relational expression**. The **relational operator** '<' converts two numbers into the logical constant TRUE. Any significant mathematical sentence that evaluates to TRUE is called a **theorem**, and expression (4.2) is one of the first theorems of analysis; it underpins all inequalities among real numbers.

Relational operators comprise very familiar objects:

$$= \quad \neq \quad < \quad \leqslant \quad > \quad \geqslant . \tag{4.3}$$

These operators are **binary**, they act on two operands. The first two act on elements of any set; the others act on real numbers (more generally, on elements of an **ordered set**—see Sect. 4.6).

Interesting things can be said with simple relational expressions:

$$9^3 + 10^3 = 1^3 + 12^3, \qquad \left| \pi - \frac{355}{113} \right| < 3 \times 10^{-7}. \tag{4.4}$$

These expressions also feature prominently in programming languages:

```
if x<1 then
    x := -x + 1
else
    x := x - 1
fi:
```

Here the logical value of the expression '$x < 1$' determines which assignment statement is executed.

At the most basic level we have the relational operators associated with sets, namely the **membership** and **inclusion operators** (Sect. 2.1) and their negation:

$$\in \quad \notin \quad \subset \quad \not\subset \quad \supset \quad \not\supset . \tag{4.5}$$

For example, the relational expression

$$\sqrt{2} \notin \mathbb{Q} \tag{4.6}$$

is TRUE, that is, the square root of 2 is not a rational number. We shall prove this fact in Sect. 7.1.

There are countless relational operators in mathematics: the **divisibility operator** '|' and the **congruence operator** '≡' in arithmetic (Sect. 2.1.3), the **isomorphism operator** '≅' in algebra, the **orthogonality operator** '⊥' in geometry, etc. By

4.2 Logical Operators

The following sentence is TRUE:

$$\text{The integer } 2^{57885161} - 1 \text{ is prime.} \tag{4.7}$$

The proof requires a sophisticated theory and a lot of computer time.[1] However, it is not difficult to see that this expression is decomposable into **finitely many** relational expressions. Indeed the primality of $p = 2^{57885161} - 1$ could in principle be established by verifying that p is not divisible by any prime up to $\sqrt{p}$. (The locution 'in principle' is cheeky; no computer present or future will ever complete such a brute-force computation within the lifetime of our Universe.)

This example motivates the introduction of the **logical** (or **boolean**) **operators** NOT and AND (to be defined below), with which we rewrite (4.7) as follows:

$$\text{NOT}\,(2|p)\quad \text{AND}\quad \text{NOT}\,(3|p)\quad \text{AND}\quad \text{NOT}\,(5|p)\quad \text{AND}\quad \cdots.$$

We have produced a complex symbolic sentence by combining relational expressions with logical operators. This process is analogous to the construction of complex arithmetical expressions from arithmetical constants (numbers) and operators $(+, -, \text{etc.})$.

The operator NOT (represented by the symbol $\neg$) is **unary**—it takes one boolean operand and produces a boolean value. In this respect NOT resembles the unary arithmetical operator '$-$', which changes the sign of a number. The operator AND (represented by the symbol $\wedge$) is **binary**: think of it as a kind of multiplication.

The following **truth tables** define the operators NOT and AND by specifying their action on all possible choices of operands:

P	$\neg P$
T	F
F	T

P	Q	$P \wedge Q$
T	T	T
T	F	F
F	T	F
F	F	F

(4.8)

The expression $\neg P$ is called the **negation** of P. Negating relational expressions is straightforward, as we already have all the relevant symbols—see (4.3) and (4.5).

---

[1] This expression refers to the largest known prime (as of January 2014) with 17425170 digits: see [6].

Thus

$$\neg(x < y) = (x \geq y), \qquad \neg(x \in A) = (x \notin A), \qquad \neg(A \not\subset B) = A \subset B.$$

The expression $P \wedge Q$ is called a **conjunction** or **compound expression**. In a compound expression the operator AND may appear implicitly. For instance, the expression $0 < x < 1$ is compound: $(0 < x) \wedge (x < 1)$.

Next we define the operator OR (represented by the symbol $\vee$) and the **implication operator** (represented by the symbol $\Rightarrow$), as follows:

$$\begin{array}{cc|c} P & Q & P \vee Q \\ \hline T & T & T \\ T & F & T \\ F & T & T \\ F & F & F \end{array} \qquad \begin{array}{cc|c} P & Q & P \Rightarrow Q \\ \hline T & T & T \\ T & F & F \\ F & T & T \\ F & F & T \end{array} \qquad (4.9)$$

These operators could also be expressed in terms of $\neg$ and $\wedge$, see Exercise 4.10. The operator OR is **inclusive**, namely $T \vee T = T$, which is not the meaning usually attributed to the conjunction 'or' in English usage.[2]

The expression

$$P \Rightarrow Q \qquad (4.10)$$

is called an **implication**, and it reads

> *P implies Q*   *Q follows from P*   *if P, then Q*
> *P only if Q*   *P is sufficient for Q.*

The expression P is the **hypothesis** (or **antecedent**) and Q is the **conclusion** (or **consequent**). Many mathematical statements have this form, for example theorem (4.1).

According to the truth table (4.9), if P is false, then the expression $P \Rightarrow Q$ is true for any Q. This convention, which is perhaps unexpected, is in fact in agreement with the common usage of an implication:

> *If 4 is a prime number, then I am a Martian.*
> *If I win the lottery, then I'll buy you a Ferrari.*

The meaning of these sentences is clear. In particular, if I don't win the lottery, then I am free do what I want about the Ferrari.

From (4.9) we see that the expressions $P \Rightarrow Q$ and $Q \Rightarrow P$ are different. Accordingly, we introduce the operator $\Leftarrow$ as follows:

$$P \Leftarrow Q \stackrel{\text{def}}{=} Q \Rightarrow P.$$

---

[2] There is also an **exclusive** version of OR, called XOR ($\veebar$), for which $T \veebar T = F$.

4.2 Logical Operators

The expression on the left is called the **converse** of the implication (4.10). It reads

P is implied by Q $\qquad$ P follows from Q
P if Q $\qquad$ P is necessary for Q.

The value of an implication and its converse are unrelated. In the following example the former is false and the latter is true:

$$(x^2 = 25) \Rightarrow (x = -5) \qquad (x^2 = 25) \Leftarrow (x = -5). \qquad (4.11)$$

Inappropriate reversal of an implication is a common mistake in proofs—see Sect. 7.7.2.

Formulae (4.11) combine relational and boolean operators, and the parentheses suggest the order in which these operators are evaluated. In this case the parentheses are redundant, since relational operators must necessarily take precedence over boolean operators. Thus the expression $x^2 = 25 \Rightarrow x = -5$ could not be interpreted as $x^2 = (25 \Rightarrow x) = -5$. However, when two or more boolean operators are present, parentheses may be inserted to change the meaning of an expression. For instance, the expressions

$$(F \wedge T) \Rightarrow T \qquad F \wedge (T \Rightarrow T)$$

evaluate to TRUE and FALSE respectively. As with arithmetic operators, there is an agreed order with which boolean operators are evaluated: first $\neg$, then $\wedge$, then $\vee$, then $\Rightarrow$. Thus the expression $P \vee \neg Q \Rightarrow R \wedge S$ is evaluated as $(P \vee (\neg Q)) \Rightarrow (R \wedge S)$; we note that the redundant parentheses add clarity to the sentence.

The **equivalence operator** $\Leftrightarrow$ is the conjunction of the direct and converse implications:

$$P \Leftrightarrow Q \stackrel{\text{def}}{=} (P \Rightarrow Q) \wedge (P \Leftarrow Q) \qquad (4.12)$$

and it corresponds to the following truth table:

P	Q	$P \Leftrightarrow Q$
T	T	T
T	F	F
F	T	F
F	F	T

The expression $P \Leftrightarrow Q$ is read aloud as

P implies and is implied by Q $\qquad$ P is equivalent to Q
P if and only if Q $\qquad$ P is necessary and sufficient for Q

The awkward expression 'if and only if' is quite common; the abbreviated form 'iff' is found in formulae as an alternative to '$\Leftrightarrow$'. So the expressions

$$A = B \Leftrightarrow A \triangle B = \emptyset \qquad\qquad A = B \text{ iff } A \triangle B = \emptyset$$

have the same meaning; we turn symbols into words:

Two sets are equal if and only if their symmetric difference is empty.

If—as in this example—the value of $P \Leftrightarrow Q$ is TRUE, then the statements P and Q are **logically equivalent**, meaning that one is is just a rewording of the other.

The **contrapositive** of the implication (4.10) is the implication

$$\neg P \Leftarrow \neg Q. \tag{4.13}$$

The contrapositive is constructed by reversing the operator and negating the operands. Great care must be exercised in distinguishing between direct, converse, and contrapositive implications.

DIRECT :	*If x is a multiple of 4, then x is even.*	(TRUE)
CONVERSE :	*If x is even, then x is a multiple of 4.*	(FALSE)
CONTRAPOSITIVE :	*If x is odd then x is not a multiple of 4.*	(TRUE)

While the value of an implication and of its converse are unrelated, every implication is equivalent to its contrapositive. This is identity (iv) of the following theorem.

Theorem 4.1 *For all $P, Q \in \{T, F\}$, the following holds:*

(i) $\quad \neg(P \vee Q) \Leftrightarrow (\neg P \wedge \neg Q)$
(ii) $\quad \neg(P \wedge Q) \Leftrightarrow (\neg P \vee \neg Q)$
(iii) $\quad (P \Rightarrow Q) \Leftrightarrow (\neg P \vee Q)$
(iv) $\quad (P \Rightarrow Q) \Leftrightarrow (\neg P \Leftarrow \neg Q).$

PROOF. We prove (iii), leaving the other cases as an exercise. We will show that the two sides of the equivalence (iii) have the same truth table. The truth table of the left-hand side is given in (4.9); that of the right-hand is computed as follows:

P	Q	$\neg P$	$\neg P \vee Q$	$P \Rightarrow Q$
T	T	F	T	T
T	F	F	F	F
F	T	T	T	T
F	F	T	T	T

We see that the two truth tables are equal. □

The statements (i) and (ii) are known as *De Morgan's laws*.[3] Using Theorem 4.1, one can express the operators $\vee, \Rightarrow, \Leftrightarrow$ in terms of $\neg$ and $\wedge$ (Exercise 4.10).

All items in the theorem have the form $A(P, Q) \Leftrightarrow B(P, Q)$, where the statements A and B have the same value for any choice of P and Q and hence are equivalent. A statement of this kind is called a **tautology**. In logic a distinction is made between a

---

[3] Augustus De Morgan (British: 1806–1871).

4.2 Logical Operators

tautology and a syntactically correct—but not necessarily true—double implication, by introducing the symbol $\leftrightarrow$ for the latter circumstance. So one would write $(P \vee Q) \leftrightarrow (P \wedge Q)$ rather than $(P \vee Q) \Leftrightarrow (P \wedge Q)$. A similar distinction exists between $\Rightarrow$ and $\rightarrow$. For our purpose these additional symbols are unnecessary.

4.3 Predicates

The sentence

The integer x is divisible by 7

contains an unspecified symbol, and thus it assumes a logical value only if the symbol is assigned an integer value. We are prompted to introduce the following function:

$$\mathscr{P} : \mathbb{Z} \to \{\text{T}, \text{F}\} \qquad x \mapsto 7 \mid x. \tag{4.14}$$

We see that $\mathscr{P}(x)$ is the symbolic translation of our sentence, and that $\mathscr{P}(22) = \text{F}$, $\mathscr{P}(-91) = \text{T}$.

Using a function to encode a sentence is a simple idea with far-reaching consequences. Let us thus define a **predicate** (or **boolean function**) to be any function that assumes boolean values. If the domain of a predicate $\mathscr{P}$ is X, then we speak of a predicate **on** (or **over**) X. So (4.14) is a predicate on the integers.

Let $\mathscr{P} : X \to \{\text{T}, \text{F}\}$ be a predicate on X. There is a distinguished subset $A_\mathscr{P}$ of X, determined by $\mathscr{P}$ via the following Zermelo definition:

$$A_\mathscr{P} := \{x \in X : \mathscr{P}(x)\}. \tag{4.15}$$

With reference to Eq. (2.9) and the discussion that follows, we see that the expression $\mathscr{P}(x)$, which means 'x has property P', is the definiens of the set $A_\mathscr{P}$. So the Zermelo definition (4.15) may be expressed in the language of functions using an inverse image:

$$\{x \in X : \mathscr{P}(x)\} \stackrel{\text{def}}{=} \mathscr{P}^{-1}(\{\text{T}\})$$

[cf. Definition (2.20)].

Conversely, let X be a set and let A be a subset of X. The predicate

$$\mathscr{P}_A : X \to \{\text{T}, \text{F}\} \qquad x \mapsto x \in A \tag{4.16}$$

is called the **characteristic function** of A (in X). Explicitly, $\mathscr{P}(x)$ is equal to T if x belongs to A and to F otherwise. For example, the function (4.14) is the characteristic function of the set $7\mathbb{Z}$ of integer multiples of 7.

So to every predicate on a set X we associate a unique subset of X, and vice-versa. We say that there is a **bi-unique correspondence** between these two classes of objects, established by expressions (4.15) and (4.16).

Sets are manipulated using set operators. Predicates are manipulated using boolean operators in a natural way by defining, say,

$$(\mathscr{P} \wedge \mathscr{Q})(x) \stackrel{\text{def}}{=} \mathscr{P}(x) \wedge \mathscr{Q}(x)$$

and similarly for all other operators. The following theorem establishes correspondences between these two classes of objects.

Theorem 4.2 *Let X be a set, let $A, B \subseteq X$, and let $\mathscr{P}_A$, $\mathscr{P}_B$ be the corresponding characteristic functions. The following holds (the prime denotes taking the complement):*

(i) $\quad \neg \mathscr{P}_A = \mathscr{P}_{A'}$
(ii) $\quad \mathscr{P}_A \wedge \mathscr{P}_B = \mathscr{P}_{A \cap B}$
(iii) $\quad \mathscr{P}_A \vee \mathscr{P}_B = \mathscr{P}_{A \cup B}$
(iv) $\quad \mathscr{P}_A \Rightarrow \mathscr{P}_B = \mathscr{P}_{(A \smallsetminus B)'}$
(v) $\quad \mathscr{P}_A \Leftrightarrow \mathscr{P}_B = \mathscr{P}_{(A \cap B) \cup (A \cup B)'}$.

(This baroque display of symbols is unappealing. In Sect. 6.6 we'll make it more digestible by writing a short essay about it without symbols.)

PROOF. To prove (i) we note that the function $x \mapsto \neg \mathscr{P}_A(x)$ evaluates to TRUE if $x \notin A$ and to FALSE otherwise. However, from the definition of the complement of a set, we have $x \notin A \Leftrightarrow x \in A'$.

Next we prove (iv). We'll prove instead

$$\neg(\mathscr{P}_A \Rightarrow \mathscr{P}_B) = \mathscr{P}_{(A \smallsetminus B)}$$

which, together with (i), gives us (iv). Let $\mathscr{P}_A := (x \in A)$ and let $\mathscr{P}_B := (x \in B)$. From the truth table (4.9) of the operator $\Rightarrow$, we find that $\neg(\mathscr{P}_A \Rightarrow \mathscr{P}_B)(x)$ is TRUE precisely when $\mathscr{P}_A(x)$ is TRUE and $\mathscr{P}_B(x)$ is FALSE. This means that

$$(x \in A) \wedge (x \notin B),$$

but this is just the definition of the characteristic function of the set $A \smallsetminus B$, as desired.

The proof of (ii), (iii), (v) is left as an exercise. □

We illustrate the significance of this theorem with examples.

EXAMPLE. Let a and b be real numbers. The predicates $x \mapsto (x \geqslant a)$ and $x \mapsto (x < b)$ (over $\mathbb{R}$) are the characteristic functions of two rays, the latter without end-point. According to Theorem 4.2 part (ii), the predicate $x \mapsto ((x \geqslant a) \wedge (x < b))$ is the characteristic function of the intersection of these rays. Depending on whether $a < b$, or $a \geqslant b$, this intersection is the half-open interval $[a, b)$ or the empty set.

EXAMPLE. Let X, Y be sets, and let $f, g : X \to Y$ be functions. An **equation** (on X) is a predicate of the type

4.3 Predicates

$$\mathscr{P} : X \to \{\text{T}, \text{F}\} \qquad x \mapsto (f(x) = g(x)).$$

In the language of predicates, an equation is the characteristic function of its own solution set $\{x \in X : \mathscr{P}(x)\}$ (Sect. 3.3). The system of two equations

$$f_1(x) = g_1(x) \qquad f_2(x) = g_2(x)$$

defined over the same set corresponds to the predicate $x \mapsto \mathscr{P}_1(x) \wedge \mathscr{P}_2(x)$. From Theorem 4.2, part (ii), it follows that the solution set of a system of two equations is the intersection of the solution sets of the individual equations. The same holds for any number of equations—see (3.19).

EXAMPLE. The standard form of the sigma-notation (see Sect. 3.2) has the following syntax:

$$\sum_{\mathscr{P}(k)} a_k \qquad (4.17)$$

where $\mathscr{P}$ is any predicate over $\mathbb{Z}$. The range of summation consists of those values of k for which $\mathscr{P}(k)$ is TRUE.

EXAMPLE. If $A \subset B$, then $A \setminus B$ is empty, and from part (iv) of the theorem we obtain

$$(\mathscr{P}_A \Rightarrow \mathscr{P}_B) = \mathscr{P}_\emptyset = \mathscr{P}_X.$$

The subset relation has been translated into an identity of functions. For example, the functional identity

$$(\mathscr{P}_{4\mathbb{Z}} \Rightarrow \mathscr{P}_{2\mathbb{Z}}) = \mathscr{P}_\mathbb{Z}$$

says that every multiple of four is even.

4.4 Quantifiers

Interesting mathematical statements—such as (4.1)—refer to families of objects rather than to individual objects. The formulation of general statements requires two special symbols, the **universal quantifier** $\forall$ and the **existential quantifier** $\exists$. These symbols translate into words in several equivalent ways:

$\forall$:   for all   given any   for any choice of
$\exists$:   for some   there exists   we can find

Expressions with quantifiers have the following syntax:

$$\begin{array}{ll} \forall x \in X, \mathscr{P}(x) & \text{For all } x \text{ in } X, \mathscr{P}(x). \\ \exists x \in X, \mathscr{P}(x) & \text{There exists } x \text{ in } X, \text{ such that } \mathscr{P}(x). \end{array} \qquad (4.18)$$

where X is a set and $\mathscr{P}$ is a predicate over X. The quantifier is followed by the symbol being quantified, the membership operator, a set, and a predicate. The acronym 's.t.' (for 'such that') is sometimes inserted in the second symbolic expression (4.18) before the predicate:
$$\exists x \in X \quad \text{s.t.} \quad \mathscr{P}(x).$$

Without a predicate the expression $\forall x \in X$ is incomplete and meaningless. The expression $\exists x \in X$ is meaningful but not used; one instead writes $X \neq \emptyset$.

The meaning of the expressions (4.18) should be clear; in particular, each expression has a logical value. For example, the following sentences are TRUE:

$\forall x \in \mathbb{R}, \ x^2 \geqslant 0$ *The square of every real number is non-negative.*
$\exists x \in \mathbb{Q}, \ 0 < x < 1$ *There is a rational number in the open unit interval.*

In translating the symbolic expressions we haven't merely converted symbols into words following (4.18). We have also synthesised meaning, and in doing so the quantified variable x has become silent—no explicit reference to it remains in the sentence. This is an intrinsic phenomenon, as we shall see.

If the ambient set is clear from the context, then reference to it may be omitted. This is invariably the case if the quantifier appears in conjunction with an inequality that restricts the variable's range. Of the following equivalent expressions

$\forall n > 3, \ n! > 2^n \qquad \forall n \in \mathbb{N}\setminus\{1,2,3\}, \ n! > 2^n \qquad \forall n \in \mathbb{N}, \ n > 3 \Rightarrow n! > 2^n$

the first one is preferable. The presence of the factorial makes it clear that we are dealing with natural numbers.

EXAMPLE. Using quantifiers, it is possible to define relational operators in terms of other operators, thereby reducing the number of primitive operators. The inclusion operator can be defined in terms of the membership operator:
$$A \subset B \overset{\text{def}}{\Leftrightarrow} \forall x \in A, \ x \in B. \tag{4.19}$$

The divisibility operator is defined in terms of multiplication:
$$m|n \overset{\text{def}}{\Leftrightarrow} \exists k \in \mathbb{Z}, \ mk = n. \tag{4.20}$$

EXAMPLE. The symbolic sentences (4.18) may be expressed in the language of functions:
$$\forall x \in X, \ \mathscr{P}(x) \Leftrightarrow \mathscr{P}(X) = \{\text{T}\}$$
$$\exists x \in X, \ \mathscr{P}(x) \Leftrightarrow \mathscr{P}(X) \neq \{\text{F}\}. \tag{4.21}$$

These equivalences may give the impression that we have managed to bypass quantifiers altogether. This is not the case. The standard definition of image of a set under

4.4 Quantifiers

a function (2.18) has a hidden quantifier in it, which is revealed when the definition is converted to Zermelo form. Thus if $f : A \to B$ is a function, then

$$f(A) = \{f(x) : x \in A\} = \{y \in B : \exists x \in A, \quad y = f(x)\}.$$

Of the two definitions, the first one is more intuitive.

Several quantifiers may appear within the same sentence. For illustration, let us consider the following two-variable predicate:

$$\mathscr{L}(x, y) := (\text{'}x \text{ loves } y\text{'}).$$

This function is defined over the cartesian product $G \times G$, where G is a set of people. So the lovers and the loved ones belong to the same set.

Now choose $g \in G$, say, $g =$ George. Then 'everybody loves George' is a boolean expression, which is spelled out as

For all elements x of G, the value of $\mathscr{L}(x, g)$ is TRUE.

This expression translates into symbols as follows:

$$\forall x \in G, \quad \mathscr{L}(x, g).$$

This sentence may be true or false (depending on how charming George is). The following sentences with one or two quantifiers exemplify several possible constructs.

	Words	Symbols
1	Everybody loves George	$\forall x \in G, \quad \mathscr{L}(x, g)$
2	Somebody loves George	$\exists x \in G, \quad \mathscr{L}(x, g)$
3	Everybody loves himself	$\forall x \in G, \quad \mathscr{L}(x, x)$
4	George is in love	$\exists x \in G \setminus \{g\}, \quad \mathscr{L}(g, x)$
5	George is selfish	$\mathscr{L}(g, g) \wedge (\forall x \in G \setminus \{g\}, \quad \neg \mathscr{L}(g, x))$
6	Everybody is in love	$\forall x \in G, \exists y \in G, \quad \mathscr{L}(x, y)$
7	Somebody is in love	$\exists x \in G, \exists y \in G, \quad \mathscr{L}(x, y)$
8	Somebody loves everybody	$\exists x \in G, \forall y \in G, \quad \mathscr{L}(x, y)$
9	Somebody is loved by everybody	$\exists y \in G, \forall x \in G, \quad \mathscr{L}(x, y)$

In the sentences 4 and 5, the indeterminate x belongs to the set $G \setminus \{g\}$, to ensure that x is distinct from g. It is possible to transfer this constraint to the predicate, e.g., in item 4:

$$\exists x \in G, \quad (x \neq g) \wedge \mathscr{L}(g, x).$$

Two-quantifier sentences have implied parentheses, which establish the order of evaluation of the sub-expressions. Thus expression 6 is written in full as

$$\forall x \in G, \quad (\exists y \in G, \quad \mathscr{L}(x, y)).$$

According to (4.18), the quantity in parentheses must be equal to $\mathscr{Q}(x)$ for some predicate $\mathscr{Q}$ over G; indeed we see that $\mathscr{Q}(x) =$ 'x loves somebody'. From the sentence's symbolic structure and meaning, it is clear that the choice of y depends on x, in general. Therefore, exchanging the order of the quantifiers alters the meaning of the expression, as confirmed by comparing sentences 6 and 9.

Quantifiers are **operators** which act on boolean functions, thereby producing new functions. Their effect is to reduce the number of variables by one, a process that calls to mind definite integration. We should think of the two expressions

$$\forall x \in X \qquad \int_X dx$$

as being structurally similar. The operator on the left acts on predicates over the set X, namely on functions $\mathscr{P} : X \to \{\text{T}, \text{F}\}$. The operator on the right acts, say, on functions $F : X \to \mathbb{R}$, which are integrable over a subset X of the real line. (To fix ideas, assume that $X = [a, b]$ is an interval.) Inserting the appropriate functions in each expression

$$\forall x \in X, \quad \mathscr{P}(x) \qquad \int_X dx\, F(x)$$

we obtain a boolean constant (TRUE or FALSE) on the left, and a numerical constant (a number) on the right. Now suppose that both $\mathscr{P}$ and F are functions of two variables x and y, defined over the cartesian product $X \times Y$ of two sets. Then each expression

$$\forall x \in X, \quad \mathscr{P}(x, y) \qquad \int_X dx\, F(x, y)$$

produces a function of y, the variable that is not quantified in one case, and not integrated over in the other. If we quantify/integrate with respect to both variables,

$$\forall x \in X, \quad \exists y \in Y, \quad \mathscr{P}(x, y) \qquad \int_X dx \int_Y dy\, F(x, y),$$

once again we obtain constants.

Returning to the examples above, we note that $\mathscr{L}(x, g)$ is a predicate in one variable (g is fixed), and so is $\mathscr{L}(x, x)$. Hence $\exists x, \mathscr{L}(x, g)$ and $\forall x, \mathscr{L}(x, x)$ are constants. On the other hand, any expression in which the number of quantifiers is smaller than the number of arguments in the predicate is a predicate with fewer arguments, as the following examples illustrate.

4.4 Quantifiers

Words	Symbols
x is in love	$\exists y \in G \setminus \{x\}, \; \mathscr{L}(x, y)$
x is loved	$\exists y \in G \setminus \{x\}, \; \mathscr{L}(y, x)$
x is a hippy	$\forall y \in G, \; \mathscr{L}(x, y)$
x is selfish	$\mathscr{L}(x, x) \wedge (\forall y \in G \setminus \{x\}, \; \neg \mathscr{L}(x, y))$
x is a lover of George	$x \neq g \wedge \mathscr{L}(x, g) \wedge \mathscr{L}(g, x)$

Now, each predicate is the characteristic function of a subset of G, which we construct with a Zermelo definition.

Words	Symbols
the people in love	$\{x \in G : \exists y \in G \setminus \{x\}, \; \mathscr{L}(x, y)\}$
the loved ones	$\{x \in G : \exists y \in G \setminus \{x\}, \; \mathscr{L}(y, x)\}$
the hippies	$\{x \in G : \forall y \in G, \; \mathscr{L}(x, y)\}$
the selfish people	$\{x \in G : \mathscr{L}(x, x) \wedge (\forall y \in G \setminus \{x\}, \; \neg \mathscr{L}(x, y))\}$
George's lovers	$\{x \in G : x \neq g \wedge \mathscr{L}(x, g) \wedge \mathscr{L}(g, x)\}$

The sentence *'everybody loves somebody'* is of the form

$$\forall x \in X, \quad \exists y \in Y, \quad \mathscr{P}(x, y) \qquad (4.22)$$

where X and Y are sets and $\mathscr{P}$ is a predicate over $X \times Y$. Letting $X = Y = \mathbb{N}$ and $\mathscr{P}(x, y) = (x < y)$, we obtain a new statement which has the same formal structure:

Given any integer, one can find a larger integer.

or better,

There is no greatest integer.

It's useful to think of the interplay between universal and existential quantifiers as a representation of an adversarial system—like a court of law—based on the following rules of engagement:

$$\forall \quad \text{given any} \quad \text{(my opponent's move)}$$
$$\exists \quad \text{I can find} \quad \text{(my move).}$$

The quantifiers create a tension between the two contenders, and I should expect to be challenged by my opponent. Accordingly, we rewrite our statement more emphatically:

Given any integer, no matter how large, one can always find a larger integer.

The expressions 'no matter how large' and 'always' are inessential to the claim, but they expose the dynamics of the process. This aspect of predicate calculus will be considered again in Chap. 7, when we deal with proofs.

EXAMPLE. The **Archimedean property** of the real numbers states that

The integer multiples of any positive quantity can be made arbitrarily large.

The symbolic version of this statement requires three quantifiers.

$$\forall x \in \mathbb{R}^+, \quad \forall y \in \mathbb{R}, \quad \exists n \in \mathbb{N}, \quad nx > y.$$

In this expression x is the positive quantity in question, y is the (large) quantity we want to exceed, and n is the multiple of x needed to achieve this. Note that $n = n(x, y)$. Clearly, the Archimedean property is better expressed with words than with symbols, and the large number of quantifiers hidden within this sentence may be taken as an indication of the concept's logical depth.

4.4.1 Quantifiers and Functions

Invariably, statements about functions will refer to the elements of the domain or co-domain, and quantifiers may be made to act on these elements. For instance, a function is **injective** if it maps distinct points to distinct points (Sect. 2.2). This concise statement unfolds as follows:

Given any two points in the domain of the function, if they are distinct, then so are their images under the function.

This sentence has a universal quantifier ('given any') and an implication operator ('if ... then'). Let $f : A \to B$. A literal translation into symbols is

$$\forall x, y \in A, \quad (x \neq y) \Rightarrow (f(x) \neq f(y))$$

where $\forall x, y \in A$ is a shorthand for $\forall x \in A, \forall y \in A$. Replacing the implication by its contrapositive [Theorem 4.1 (iv)] we obtain a neater expression:

$$\forall x, y \in A, \quad (f(x) = f(y)) \Rightarrow (x = y).$$

The original definition ('distinct points have distinct images') restricts the ambient set to pairs of distinct points. This can be done symbolically:

$$\forall x \in A, \quad \forall y \in A \setminus \{x\}, \quad f(x) \neq f(y).$$

Now the predicate is simpler but the set is more complicated. It is always possible to transform an implication by absorbing the hypothesis into the ambient set (Exercise 4.10.3).

A function f is **surjective** if image and co-domain coincide: $f(A) = B$. This high-level expression hides two quantifiers; we first spell it out:

4.4 Quantifiers

Given any point in the co-domain, we can find a point in the domain which maps to it.

and then we rewrite it symbolically:

$$\forall y \in B, \quad \exists x \in A, \quad f(x) = y.$$

This expression is of the form (4.22).

In Sect. 3.1 we noted that a sequence of elements of a set A may be interpreted as a function:

$$a : \mathbb{N} \to A \quad k \mapsto a_k.$$

Characterising sequences is then analogous to characterising functions.

For example, consider the set $\mathbb{Z}^\mathbb{N}$ of all integer sequences (this notation was developed in Sect. 3.5.1). To isolate sequences with certain properties—a subset of $\mathbb{Z}^\mathbb{N}$—we use a Zermelo definition. For example, the set

$$\{a \in \mathbb{Z}^\mathbb{N} : a_1 \in 2\mathbb{Z}\}$$

consists of all integer sequences whose first term is even. By applying the quantifier $\forall$ to the subscript k (which is the variable in our functions), we can deal with all terms of the sequences at once. So the set of integer sequences with only even terms is given by

$$\{a \in \mathbb{Z}^\mathbb{N} : \forall k \in \mathbb{N}, \ a_k \in 2\mathbb{Z}\} = (2\mathbb{Z})^\mathbb{N}. \tag{4.23}$$

There seems to be something wrong here. The quantifier 'integrates out' the indeterminate k: does this mean that the predicate is just a constant? No, because the Zermelo variable is actually $a = (a_k)$, an element of the ambient set $\mathbb{Z}^\mathbb{N}$. If we write (4.23) as $\{a \in \mathbb{Z}^\mathbb{N} : \mathscr{P}(a)\}$, then we see the structure of the predicate:

$$\mathscr{P} : \mathbb{Z}^\mathbb{N} \to \{\text{T}, \text{F}\} \quad a \mapsto (\forall k \in \mathbb{N}, \ a_k \in 2\mathbb{Z}).$$

Likewise, the set

$$\{a \in \mathbb{Z}^\mathbb{N} : \exists k \in \mathbb{N}, \ a_k \in 2\mathbb{Z}\}$$

is

The set of integer sequences with an even term.

The above expression reflects the very literal interpretation of what's written, which is what mathematicians—and lawyers—are notorious for. A mathematical geek would be entertained by the fact that the statement '*In London there is an underground station*', is true; a normal person would instead perceive this as a puzzling understatement ('Where do they go from there?'). So a characterisation of the type

The set of integer sequences with at least one even term.

is preferable to the one given before. The qualifier 'at least' is superfluous, but helps the reader note an essential point.

The set
$$\{a \in \mathbb{Z}^{\mathbb{N}} : \forall n \in \mathbb{N}, \ n > 2 \Rightarrow 2|a_n\}$$

is described as

The set of integer sequences whose terms, after the second, are even.

No information is given about the first two terms, yet this statement could be misinterpreted as meaning that these terms are odd. A more eloquent description of this set is

The set of integer sequences whose terms are all even, with the possible exception of the first two terms.

These examples show how much can be done to improve the clarity of a definition; this will be our concern in Chap. 6.

4.4.2 Existence Statements

An **existence statement** is a logical expression with a leading existential quantifier $\exists$. In this section we consider some examples; we shall take a deeper look at the question of existence in Chap. 9.

The standard symbolic form of an existence statement is (4.18), but in a mathematical sentence the quantifier is often hidden:

The number $\cos(\pi/3)$ is rational.

To make the quantifier visible, we write

For some rational number r, we have $\cos(\pi/3) = r$.

or, in symbols,

$$\exists r \in \mathbb{Q}, \ r = \cos(\pi/3).$$

This statement is weaker than the identity $\cos(\pi/3) = 1/2$, because it does not require us to reveal which rational number our expression is equal to.

Equation (4.20) shows that an existential quantifier is hidden in the definition of the divisibility of integers. Therefore the sentence 'n is even', or, more generally, 'm divides n' are existence statements.

The sentence 'n is not prime' is an existence statement, because it means '*there is a proper divisor of n*'. We write it symbolically as

$$\exists m \in \mathbb{N}, \ m \notin \{1, n\} \wedge (m|n)$$
$$\Leftrightarrow \ \exists m \in \mathbb{N}, \ m \notin \{1, n\} \wedge (\exists k \in \mathbb{N}, \ mk = n)$$

4.4 Quantifiers

where we have been careful in excluding trivial divisors.

The following well-known theorem in arithmetic is an existence statement.

Theorem (Lagrange,[4] 1770). *Every natural number is the sum of four squares.*

Let us analyse it in detail. First, we consider two instances of the theorem:

$$5 = 2^2 + 1^2 + 0^2 + 0^2 \qquad 7 = 2^2 + 1^2 + 1^2 + 1^2.$$

The first identity shows that some integers are the sum of fewer than four squares, while one checks that 7 requires all four squares. We restate the second identity without disclosing the details:

$$\exists a, b, c, d \in \mathbb{Z}, \ 7 = a^2 + b^2 + c^2 + d^2.$$

By replacing 5 or 7 with an unspecified natural number n, we obtain a predicate $\mathscr{L}$ over $\mathbb{N}$:

$$\mathscr{L}(n) := \left(\exists a, b, c, d \in \mathbb{Z}, \ n = a^2 + b^2 + c^2 + d^2 \right) \qquad (4.24)$$

or

n is a sum of four squares.

To state Lagrange's theorem with symbols, we quantify the remaining variable n:

$$\forall n \in \mathbb{N}, \ \exists a, b, c, d \in \mathbb{Z}, \ n = a^2 + b^2 + c^2 + d^2. \qquad (4.25)$$

Five quantifiers are necessary to turn the predicate $n = a^2 + b^2 + c^2 + d^2$ (a function of five variables) into a boolean constant. In the original formulation of this theorem, all five variables have fallen silent.

Having buried all meaning inside the predicate (4.25), Lagrange's theorem now disappears into a set identity:

$$\mathbb{N} = \{ n \in \mathbb{N} : \mathscr{L}(n) \}.$$

Any theorem on natural numbers can be put in this form for an appropriate predicate $\mathscr{L}$. (Think about it.)

Negating expressions with a leading universal quantifier always results in existence statements, as we shall see in the next section.

4.5 Negating Logical Expressions

If $\mathscr{L}$ is a logical expression, then its negation is $\neg \mathscr{L}$, which is false if $\mathscr{L}$ is true and vice-versa. Relational expressions are negated by crossing out the operator with a

---

[4] Joseph-Louis Lagrange, born Giuseppe Lodovico Lagrangia (Italian: 1736–1813).

forward slash—see (4.3) and (4.5), Sect. 4.1. The only exceptions are the inequalities, whose negations have dedicated symbols. □

Theorem 4.1 provides the negation formulae for the main compound expressions

$$\neg(P \wedge Q) \Leftrightarrow \neg P \vee \neg Q$$
$$\neg(P \vee Q) \Leftrightarrow \neg P \wedge \neg Q \qquad (4.26)$$
$$\neg(P \Rightarrow Q) \Leftrightarrow P \wedge \neg Q.$$

The first two formulae are items (i) and (ii) of the theorem (de Morgan's laws); the third follows immediately from (iii) and (i). Note that the negation of an implication does not have the form of an implication.

If $\mathscr{L}$ is an expression with quantifiers and boolean operators, then the expression $\neg\mathscr{L}$ unfolds into an equivalent expression which we seek to determine. The following sentences contain a negation and a universal quantifier:

> *Not all polynomials have real roots. A square matrix is not necessarily invertible.*

We recognise that these are disguised existence statements:

> *There is a polynomial with no real roots. There is a square matrix which is not invertible.*

The following theorem confirms this observation.

Theorem 4.3 *Let $\mathscr{P}$ be a predicate on a set A, and let*

$$\mathscr{L} := \forall x \in A, \ \mathscr{P}(x) \qquad \mathscr{M} := \exists x \in A, \ \mathscr{P}(x).$$

Then

$$\neg\mathscr{L} \Leftrightarrow \exists x \in A, \ \neg\mathscr{P}(x) \qquad \neg\mathscr{M} \Leftrightarrow \forall x \in A, \ \neg\mathscr{P}(x). \qquad (4.27)$$

PROOF. From (4.21), we have the boolean equivalence

$$\mathscr{L} \Leftrightarrow \mathscr{P}(A) = \{\text{T}\}$$

where $\mathscr{P}(A)$ is the image of A under $\mathscr{P}$ (see Sect. 2.2). But then

$$\neg\mathscr{L} \Leftrightarrow \mathscr{P}(A) \neq \{\text{T}\} \Leftrightarrow \neg\mathscr{P}(A) \neq \{\text{F}\}.$$

Using again (4.21), we obtain

$$\neg\mathscr{L} \Leftrightarrow \exists x \in A, \ \neg\mathscr{P}(x)$$

as claimed.

The second formula is proved similarly—see exercises. □

4.5 Negating Logical Expressions

It's now easy to deduce the rule for negating expressions with two or more quantifiers. Let $\mathscr{P}$ be a predicate on $A \times B$, and let us consider the expression

$$\mathscr{L} := \forall x \in A, \; \exists y \in B, \; \mathscr{P}(x, y).$$

Repeated applications of Theorem 4.3 give

$$\neg\mathscr{L} \Leftrightarrow \neg\big(\forall x \in A, \; (\exists y \in B, \; \mathscr{P}(x, y))\big)$$
$$\Leftrightarrow \exists x \in A, \; \neg(\exists y \in B, \; \mathscr{P}(x, y))$$
$$\Leftrightarrow \exists x \in A, \; \forall y \in B, \; \neg\mathscr{P}(x, y).$$

It should be clear how similar formulae may be derived, having any combination of two quantifiers.

Finally, we consider the case of an arbitrary sequence of quantifiers, e.g.,

$$\mathscr{L} = \forall x_1 \in X_1, \; \exists x_2 \in X_2, \ldots, \forall x_n \in X_n, \; \mathscr{P}(x_1, \ldots, x_n)$$

where the X_i are sets, and $\mathscr{P}$ is a predicate on the cartesian product $X_1 \times X_2 \times \cdots \times X_n$. Parentheses make it clear that this is a nested array of predicates:

$$\mathscr{L} = \forall x_1 \in X_1, \; (\exists x_2 \in X_2, \; (\ldots, (\forall x_n \in X_n, \; \mathscr{P}(x_1, \ldots, x_n)))).$$

Repeatedly applying Theorem 4.3, from the outside to the inside, we obtain

$$\neg\mathscr{L} \Leftrightarrow \exists x_1 \in X_1, \; \forall x_2 \in X_2, \ldots, \exists x_n \in X_n, \; \neg\mathscr{P}(x_1, \ldots, x_n).$$

Namely, the negation of $\mathscr{L}$ is obtained by replacing each $\forall$ with $\exists$, and vice-versa, *without changing their order*, and then replacing $\mathscr{P}$ with $\neg\mathscr{P}$. A formal proof requires the principle of induction (see Chap. 8).

EXAMPLE. The statement

$$\forall n, a, b \in \mathbb{Z}, \; (n|ab) \Rightarrow (n|a \vee n|b) \tag{4.28}$$

is false (why?). We negate it using Theorem 4.3, obtaining a true statement:

$$\exists n, a, b \in \mathbb{Z}, \; (n|ab) \wedge (n \nmid a \wedge n \nmid b).$$

EXAMPLE. Lagrange's theorem, expressed symbolically in Eq. (4.25), would be false if four squares were replaced by three squares. This fact is written symbolically as

$$\exists n \in \mathbb{N}, \; \forall a, b, c \in \mathbb{Z}, \; n \neq a^2 + b^2 + c^2.$$

In words:

There is a natural number which cannot be written as the sum of three squares.

4.6 Relations

(This section is not essential for the rest of the book: it may be skipped on first reading.)

Relations are special types of logical functions which are ubiquitous in higher mathematics. We introduce some relevant words and symbols.

Let X and Y be sets. A **relation** $\mathscr{R}$ on $X \times Y$ is a predicate over $X \times Y$. If $X = Y$, we speak of a **relation on** X. It is customary to write $x\mathscr{R}y$ to mean $\mathscr{R}(x, y)$; the expression $x\mathscr{R}y$ is called a **relational expression**, and $\mathscr{R}$ is a **relational operator**. All relational expressions introduced at the beginning of this chapter are of this form. Thus the membership operator $\in$ defines a relation on $X \times \mathbf{P}(X)$, where X is some ambient set.

A relation on a set X is sometimes defined as a **subset** of X^2, rather than a predicate over X^2. In this sense, the set $\{(1, 1), (2, 1)\}$ is a relation on $\{1, 2\}$. The correspondence between sets and predicates described in Sect. 4.3 clarifies the connection between the two constructs.

Among relations on a set X, the **equivalence relations** hold a special place. They are defined by the following properties:

$$\forall x \in X, \quad x\mathscr{R}x \qquad \text{reflexivity}$$
$$\forall x, y \in X, \quad x\mathscr{R}y \Rightarrow y\mathscr{R}x \qquad \text{symmetry}$$
$$\forall x, y, z \in X, \quad (x\mathscr{R}y \wedge y\mathscr{R}z) \Rightarrow x\mathscr{R}z \qquad \text{transitivity}.$$

The notation $x \sim y$ is usually employed to denote equivalence.

Equivalence relations allow us to regroup the elements of a set according to certain criteria. Specifically, given an equivalence relation on X and $x \in X$, one collects together in the set $[x]$ all elements of X equivalent to x:

$$[x] := \{y \in X : x \sim y\}.$$

The collection of such **equivalence classes** forms a **partition** of X (see Sect. 2.3.1), given by:

$$X/\!\sim \; := \{[x] : x \in X\}. \tag{4.29}$$

The set $X/\!\sim$ is called the **quotient set** of X by $\sim$; it's a set of sets. The idiomatic notation $X/\!\sim$ reminds us of the way this set is constructed: we subdivide X according to the criterion '$\sim$'. The minimalist definition (4.29) is sleek but inefficient, since all the elements of X that belong to the same equivalence class result in just one partition element. So there is a wholesale repetition of partition elements, and the formula works because we agreed to identify repeated elements in set definitions.

The following examples illustrate some applications of this construct.

EXAMPLE. The relational operator '$=$' defines an equivalence relation on any set. This is the **trivial** equivalence; it corresponds to the trivial partition consisting of one-element subsets.

4.6 Relations

EXAMPLE. Given a function $f : X \to Y$, we let $x_1 \sim x_2$ if $f(x_1) = f(x_2)$. This is an equivalence relation on X. The equivalence classes are the inverse images of the elements of the co-domain of f:

$$\{f^{-1}(\{y\}) : y \in Y\}. \qquad (4.30)$$

Any equivalence relation on a set X can be represented in this way, for some function f.

EXAMPLE. For any natural number m, the congruence relation $x \equiv y \pmod{m}$ defined in Sect. 2.1.3 is an equivalence relation on $\mathbb{Z}$. The equivalence classes are the congruence classes introduced in Sect. 2.3.2.

EXAMPLE. The relation on $\mathbb{N} \times \mathbb{N}$, defined by $(m, n) \sim (j, k)$ if $m + k = n + j$, is an equivalence relation. By interpreting the pair (m, n) as the quantity $z = m - n$, we see that equivalent pairs correspond to the same value of z. With this device, one can construct integers from pairs of natural numbers, whereby every integer is an equivalence class of infinitely many pairs of natural numbers. The value of this abstract construction lies with its reductionist character: it requires only natural numbers and addition in $\mathbb{N}$; there is no mention of negative integers or subtraction.

EXAMPLE. The relation '$\sim$' on $\mathbb{Z} \times (\mathbb{Z}\setminus\{0\})$, defined by $(m, n) \sim (j, k)$ if $mk = nj$, is an equivalence relation. By interpreting the pair (m, n) as $r = m/n$, we see that equivalent pairs correspond to the same value of r. With this device, one can define a rational number as an infinite collection of equivalent pairs of integers, without introducing fractions or division.

A relation $\mathscr{R}$ on a set X is called a **partial ordering** if it is reflexive, transitive and **anti-symmetric**. The latter property is defined as follows:

$$\forall x, y \in X, \ (x \mathscr{R} y \wedge y \mathscr{R} x) \Rightarrow x = y \qquad \textbf{anti-symmetry}$$

A set is **partially ordered** if a partial ordering is defined on it. A partial ordering is usually denoted by the symbol '$\leqslant$'. So we write $x \leqslant y$ instead of $x\mathscr{R}y$.

The relational operator $\leqslant$ defines a partial ordering in $\mathbb{N}, \mathbb{Z}, \mathbb{Q}$, and $\mathbb{R}$ (but not in $\mathbb{C}$). The set $\mathbf{P}(X)$ of all subsets of a set X is partially ordered by set inclusion, whereby $\leqslant$ means $\subset$.

A partially ordered set X is said to be **ordered** if all pairs of elements of X are **comparable**, meaning that we either have $x \leqslant y$ or $y \leqslant x$. The real line is an ordered set; the power set $\mathbf{P}(X)$ of a set X, which is partially ordered by set inclusion, is not ordered.

An ordered set X is said to be **well-ordered** if any non-empty subset $A \subset X$ has a smallest element. The symbolic definition requires three quantifiers:

$$\forall A \in \mathbf{P}(X)\setminus\emptyset, \ \exists a \in A, \ \forall x \in A, \ a \leqslant x.$$

Any finite ordered set is well-ordered. The closed unit interval $[0, 1]$ is ordered but not well-ordered, because the subset $(0, 1]$ has no smallest element. The natural numbers are well-ordered. This property forms the basis of the principle of induction, which we consider in Chap. 8.

Exercise 4.1 Rewrite each symbolic sentence using the quantifier $\exists$.

1. $Y \not\supset X$
2. $X \cap Y \neq \emptyset$
3. $\#X > 1$
4. $z \in X \times Y$
5. $X \neq Y$
6. $\#(X \triangle Y) > 0$
7. $\sin \circ \cos \neq \cos \circ \sin$.

Exercise 4.2 Write each sentence with symbols, using at least one quantifier.

1. The integer n is a cube.
2. The fraction a/b is not reduced.
3. The equation $f(x) = 0$ has a rational solution.
4. The unit circle has a rational point.
5. The polynomial $p(x)$ has no integer roots.
6. The function $f : A \to B$ is not injective.
7. The function $f : A \to B$ is not surjective.
8. The function $f : A \to B$ is constant.
9. The function $f : A \to B$ is not constant.
10. Every cubic equation with real coefficients has a real solution.
11. The integer n has a square divisor.[5]
12. The integer n is not divisible by 3.
13. Some integers can be written in two different ways as the sum of two cubes.[6]
14. Any open interval contains a rational number.
15. No open interval contains a smallest element.
16. The equation $f(x) = 0$ has infinitely many integer solutions.
17. The equation $f(x) = 0$ has finitely many rational solutions.

Exercise 4.3 Consider the following statements.

1. Every prime greater than 2 is odd.
2. Two integers which are co-prime are also prime.
3. The reciprocal of a positive reduced fraction is reduced.
4. The product of two primes has exactly four divisors.
5. Every divisor of the product of two integers also divides one of the factors.
6. Every subset of an infinite set is infinite.
7. Three consecutive odd integers greater than 3 cannot all be prime.
8. Two complex numbers whose sum is a real number must be complex conjugate of each other.
9. The inverse of an invertible matrix is invertible.

---

[5] The trivial divisor 1^2 is excluded.
[6] See (4.4).

4.6 Relations

In each case:

(i) Decide if the statement is true or false, hence rewrite it as an explicit implication (*For all …x, if x …, then …*).
(ii) State the contrapositive.
(iii) State the converse, and decide whether it's true or false.
(iv) State the negation.

Exercise 4.4 (W. Hodges). We wish to to build up a set of predicates to describe family relations. You are given the two predicates

$$x \text{ is a son of } y \qquad x \text{ is a daughter of } y.$$

Your task is to write definitions of the following predicates, in some appropriate order such that the later definitions use only the given predicates and earlier definitions. (The order below is just alphabetical.)

x is an aunt of y
x is a brother of y
x is a child of y
x is the father of y
x is female
x is a grandchild of y
x is a half-brother of y
x is male
x is the mother of y
x is a nephew of y
x is a parent of y
x is a sister of y.

Use the symbols x, y, z, etc., to denote people, words for everything else, and parentheses to specify the order of evaluation of logical operators. Why are the given data not quite sufficient to define the 'aunt' predicate?

Exercise 4.5 Let x and y be natural numbers, and let $\mathscr{P}(x, y)$ mean: 'x is a proper divisor of y' (that is, $x|y$ and $x \neq 1, y$). Thus $\mathscr{P}$ is a predicate over $\mathbb{N} \times \mathbb{N}$.

(a) Write each statement with words. [✦]

1. $\exists x \in \mathbb{N}, \quad \mathscr{P}(x, 5)$
2. $\exists x \in \mathbb{N}, \quad \mathscr{P}(5, x)$
3. $\forall x \in \mathbb{N}, \quad \mathscr{P}(x, x^2)$
4. $\exists x \in \mathbb{N}, \quad \forall y \in \mathbb{N}, \quad \mathscr{P}(x, y)$
5. $\forall x \in \mathbb{N} \setminus \{1\}, \quad \exists y \in \mathbb{N}, \quad \mathscr{P}(x, y).$

(b) Write each predicate with words, apart from the predicate's argument. [✎]

1. $y \mapsto (\exists x \in \mathbb{N}, \;\mathscr{P}(x, y))$
2. $y \mapsto (\forall x \in \mathbb{N}, \;\neg\mathscr{P}(x, y))$
3. $x \mapsto (\exists y \in \mathbb{N}, \;\mathscr{P}(x, y))$.

(c) Define each set with words. [✎]

1. $\{y \in \mathbb{N} : \exists x \in \mathbb{N}, \;\mathscr{P}(x, y)\}$
2. $\{y \in \mathbb{N} : \forall x \in \mathbb{N}, \;\neg\mathscr{P}(x, y)\}$
3. $\{x \in \mathbb{N} : \exists y \in \mathbb{N}, \;\mathscr{P}(x, y)\}$.

Exercise 4.6 The following expressions define sets; turn symbols into words. [✎]

1. $\{x \in \mathbb{N} : \forall m, n \in \mathbb{N}, \;(x|mn) \Rightarrow (x|m \vee x|n)\}$
2. $\{x \in \mathbb{N} : \forall m \in \mathbb{N}, \;x|m^2 \Rightarrow x|m\}$
3. $\{x \in \mathbb{N} : \forall m \in \mathbb{N}, \;x|m^3 \Rightarrow x|m^2\}$.

Exercise 4.7 Decide if each sentence is true or false, hence write its negation with symbols. Then rewrite both with words. [✎]

1. $\forall n \in \mathbb{N}, \;1/n \notin \mathbb{N}$
2. $\forall n \in \mathbb{N}, \;\sqrt{n} \in \mathbb{R} \setminus \mathbb{Q}$
3. $\forall x, y \in \mathbb{R}, \;xy = yx$
4. $\forall n \in \mathbb{Z}, \;2 | n(n+1)$
5. $\forall m, n \in \mathbb{Z}, \;(m+n \in 1+2\mathbb{Z}) \Rightarrow (m \in 2\mathbb{Z} \vee n \in 2\mathbb{Z})$
6. $\forall n \in \mathbb{N}, \;\exists r \in \mathbb{Q}, \;n < r^2 < n+1$
7. $\forall y \in \mathbb{R}, \;\exists x \in \mathbb{R}^+, \;\log(x) = y$
8. $\forall x, y > 1, \;\exists n \in \mathbb{N}, \;x^n > y$
9. $\forall \epsilon > 0, \;\exists r \in \mathbb{Q}, \;|r - \sqrt{2}| < \epsilon$.

Exercise 4.8 The following cryptic symbolic sentence states an interesting mathematical fact:

$$\forall n \in \mathbb{N}, \;\exists k \in \mathbb{N}, \;\forall j \in \mathbb{N}, \;(k+j)! \geq n^{k+j}. \tag{4.31}$$

Rewrite it in words.

Exercise 4.9 Consider the sentences

1. *Everybody loves George*
2. *Everybody loves somebody*
3. *Somebody loves everybody.*

We wish to determine the computational effort required to find out if a sentence is true or false. With the notation of Sect. 4.4, let the set G have n elements. First suppose that the sentence is true. How many evaluations of the predicate 'x loves y' are needed to verify this? Give the minimum and maximum number of evaluations, with an explanation. Do the same assuming that the sentence is false.

4.6 Relations

Exercise 4.10 In this exercise we collect some straightforward proofs.

1. Show that the operators $\vee, \Rightarrow, \Leftrightarrow$ may be defined in terms of $\neg$ and $\wedge$.
2. Complete the proof of Theorems 4.1, 4.2, and 4.3.
3. Show that given any predicates $\mathscr{P}, \mathscr{Q}$ over a set X, there is a subset A of X such that
$$(\forall x \in X, \ \mathscr{P}(x) \Rightarrow \mathscr{Q}(x)) \ \Leftrightarrow \ (\forall x \in A, \ \mathscr{Q}(x)).$$

This equivalence says that the hypothesis of an implication may be absorbed into a suitable ambient set.

Exercise 4.11 Define some interesting predicates on the power set of $\mathbb{Z}$. Do the same for the power set of $\mathbb{Z}[x]$.

Exercise 4.12 Let X be any set. Show that there is a one-to-one correspondence between the partitions of X and the equivalence relations on X. Show that any equivalence relation on X may be represented by a suitable function $f : X \to Y$ such that the equivalence classes are the pre-images of the elements of Y—see Eq. (4.30).

Exercise 4.13 You are given an equivalence relation over a finite set X. Develop an algorithm—more efficient than formula (4.29)—to construct the associated partition of X.

Chapter 5
Describing Functions

We know only a handful of words to describe properties of functions:

> *injective, surjective, invertible (bijective), constant.*

In this chapter we consider attributes of **real functions** $f : \mathbb{R} \to \mathbb{R}$. We expand our vocabulary substantially, and introduce important terms that are found in more general settings (*periodic, bounded, continuous*). Then we export this terminology to the **real sequences**, which are functions defined over $\mathbb{N}$.

5.1 Ordering Properties

The real line $\mathbb{R}$ is **ordered**, meaning that for any pair (x, y) of real numbers, precisely one of the three relational expressions

$$x < y \qquad x = y \qquad x > y$$

is true, and the other two are false. We begin with properties that are formulated in terms of ordering.

A simple attribute of a real function is the sign of the values it assumes.

$$\forall x \in \mathbb{R}, \quad f(x) > 0 \qquad \textbf{positive} \qquad (5.1)$$
$$\forall x \in \mathbb{R}, \quad f(x) < 0 \qquad \textbf{negative} \qquad (5.2)$$

If (5.1) is formulated with the non-strict inequality $f(x) \geqslant 0$, then we say that the function is **non-negative**. For the inequality $f(x) \leqslant 0$, the term 'non-positive' is uncommon, and one would normally use **negative or zero**. For example, the exponential function is positive, the absolute value function is non-negative, and the sine function is neither positive nor negative. Because inequalities are reversed under sign change, if a function f has any of the stated properties, then $-f$ has the complementary property (e.g., if f is positive, then $-f$ is negative).

© Springer-Verlag London 2014
F. Vivaldi, *Mathematical Writing*, Springer Undergraduate Mathematics Series,
DOI 10.1007/978-1-4471-6527-9_5

Note that the terms 'non-negative' and 'not negative' have different meaning, the latter being the logical negation of negative. This distinction is very clear in the symbolic definitions.

$$\forall x \in \mathbb{R}, \quad f(x) \geqslant 0 \qquad \textbf{non-negative}$$
$$\exists x \in \mathbb{R}, \quad f(x) \geqslant 0 \qquad \textbf{not negative}$$

(The second symbolic expression is obtained by negating (5.2) according to Theorem 4.3.) So a non-negative function is also not negative, but not vice-versa. Two similar expressions with different meanings can easily lead to confusion, and one must remain vigilant.

Next we consider how the action of a function affects the ordering of the real line; the order may be preserved, reversed, or a bit of both. There are two competing terminologies, labelled *I* and *II* in the table below. Each has advantages and disadvantages.

	I	*II*
$\forall x, y \in \mathbb{R}, \; x > y \Rightarrow f(x) > f(y)$	**increasing**	**strictly increasing**
$\forall x, y \in \mathbb{R}, \; x > y \Rightarrow f(x) \geqslant f(y)$	**non-decreasing**	**increasing**
$\forall x, y \in \mathbb{R}, \; x > y \Rightarrow f(x) < f(y)$	**decreasing**	**strictly decreasing**
$\forall x, y \in \mathbb{R}, \; x > y \Rightarrow f(x) \leqslant f(y)$	**non-increasing**	**decreasing**

A function that is either increasing or decreasing (strictly or otherwise) is said to be **monotonic**.

Thus the arc-tangent is increasing in *I* and strictly increasing in *II*. In *I* no function can be both increasing and decreasing, and most functions are neither, for instance a constant or the cosine. The constants are the functions that are both non-decreasing and non-increasing.

The disadvantage of *I* is that, as we did above for 'non-negative', we must differentiate between 'non-increasing' and the logical negation of increasing ($\exists x, y \in \mathbb{R}, (x > y) \land (f(x) \leqslant f(y))$).

The terminology *II* eliminates the annoying distinction between the prefixes 'non-' and 'not', but introduces a new problem. Now a constant is both increasing and decreasing, clashing with common usage. (After years without a pay rise, I wouldn't say: 'my salary is increasing'.)

To be safe, use 'strictly' to mean strict inequality in any case.

5.2 Symmetries

Real functions may have symmetries, expressing invariance with respect to changes of the argument. A function f is **even** or **odd**, respectively, if

$$\forall x \in \mathbb{R}, \quad f(-x) = f(x) \qquad \text{or} \qquad \forall x \in \mathbb{R}, \quad f(-x) = -f(x).$$

Fig. 5.1 An odd function

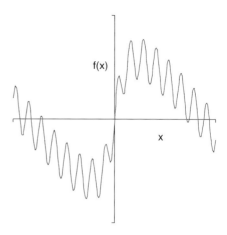

So the cosine is even, the sine is odd, the exponential is neither even nor odd, and the zero function is both even and odd. The property of being even or odd has a geometrical meaning: graphs of even functions have a mirror symmetry with respect to the ordinate axis, while those of odd functions are symmetrical with respect to the origin (see Fig. 5.1). There is an easy way of constructing even/odd functions. For any real function g, the function $x \mapsto g(x) + g(-x)$ is even, and so is the function $g(f(x))$ for any even function f. So the function $x \mapsto g(|x|)$ is even. To construct odd functions, we first define the **sign function**:

$$\text{sign}(x) = \begin{cases} +1 & \text{if } x > 0 \\ 0 & \text{if } x = 0 \\ -1 & \text{if } x < 0 \end{cases} \quad x \in \mathbb{R}. \tag{5.3}$$

The sign function is odd: $\text{sign}(-x) = -\text{sign}(x)$. Then, for any real function g, the function $x \mapsto \text{sign}(x) g(|x|)$ is odd, as easily verified. This construct ensures that our function vanishes at the origin, which is a property of all odd functions.

Next we turn to translational symmetry, which is called **periodicity**. A function f is **periodic with period** T if

$$\forall x \in \mathbb{R}, \quad f(x + T) = f(x) \tag{5.4}$$

for some non-zero real number T (see Fig. 5.2).

For instance, the sine function is periodic with period $T = 2\pi$. If a function is periodic with period T, then it is also periodic with period $2T$, $3T$, etc. For this reason one normally requires the period T to be the smallest positive real number for which (5.4) is satisfied. To emphasise this point we use the terms **minimal** or **fundamental period**.

If we say that a function f is **periodic**—without reference to a specific period—then the existence of the period must be required explicitly:

Fig. 5.2 A periodic function

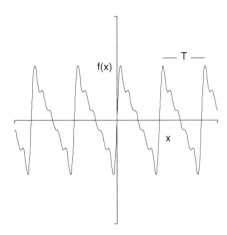

$$\exists T \in \mathbb{R}^+, \quad \forall x \in \mathbb{R}, \quad f(x+T) = f(x) \tag{5.5}$$

where $\mathbb{R}^+$ is the set of positive real numbers—see (2.12). Note that the period T must be non-zero (lest this definition say nothing) and **without loss of generality**[1] we shall require the period to be positive. [The case of negative period is dealt with the substitution $x \mapsto x - T$ in (5.4).]

The presence of symmetries reduces the amount of information needed to specify a function. If a function is even or odd, knowledge of the function for non-negative values of the argument suffices to characterise it completely. Likewise, if a periodic function is known over any interval of length equal to the period, then the function is specified completely.

One should be aware that symmetries are special; a function chosen 'at random' will have no symmetry.

5.3 Boundedness

A set $X \subset \mathbb{R}$ is **bounded** if there is an interval containing it,[2] namely if

$$\exists a, b \in \mathbb{R}, \quad \forall x \in X, \quad a < x < b \tag{5.6}$$

or

$$\exists c \in \mathbb{R}, \quad \forall x \in X, \quad |x| < c. \tag{5.7}$$

---

[1] This expression indicates that an inessential restriction or simplification is being introduced. See Sect. 7.8 for another example.

[2] The term 'interval' is intended in the proper sense—infinite intervals are excluded, cf. (2.13).

5.3 Boundedness

The two definitions are equivalent. (Think about it.) The numbers a and b in (5.6) are an **upper** and a **lower bound** for X.

A real function f is **bounded** if its image $f(\mathbb{R})$ is a bounded set. In symbols:

$$\exists c \in \mathbb{R}, \ \forall x \in \mathbb{R}, \ |f(x)| < c.$$

For example, the sine function is bounded and the exponential is not. The periodic function displayed in Fig. 5.2 is bounded.

A function f is **bounded away from zero** if its reciprocal is bounded. This means that for some positive constant c we have $|f(x)| > c$ for all values of x. In symbols:

$$\exists c \in \mathbb{R}^+, \ \forall x \in \mathbb{R}, \ |f(x)| > c.$$

The hyperbolic cosine is bounded away from zero (what could be a value of c in this case?) but the exponential function is not.

5.4 Neighbourhoods

A **neighbourhood** of a point $x \in \mathbb{R}$ is any *open* interval containing x. Although this definition makes no reference to the size of the interval, a neighbourhood of x contains all points sufficiently close to x. Thus the neighbourhood concept characterises 'proximity' in a concise manner that does not require quantitative information. A skillful use of this term leads to terse and incisive statements. For instance, the sentence

The function f is bounded in a neighbourhood of x_0.

means that there is an open interval containing x_0 whose image under f is a bounded set. If we write this statement in symbols

$$\exists a, b \in \mathbb{R}, \ \exists c \in \mathbb{R}^+, \ x_0 \in (a, b) \wedge (\forall x \in (a, b), \ |f(x)| < c)$$

we realise just how much information is packed into it. The following variant of the sentence above:

The function f is bounded in a sufficiently small neighbourhood of x_0.

says exactly the same thing, but more eloquently; the reader is warned that the required neighbourhood may be very small.

A property of a function—boundedness in this case—which holds in a neighbourhood of some point of the domain of the function, but not necessarily in the whole domain, is said to be **local**. So a function may be **locally increasing**, **locally injective**, etc.

The function f has a **maximum** at x if the value of f at x is greater than the value at all other points, namely if

$$\forall y \in \mathbb{R}\setminus\{x\}, \quad f(x) > f(y). \tag{5.8}$$

The function f has a **local maximum** at x if (5.8) holds in some neighbourhood of x. The concept of **minimum** and **local minimum** are defined similarly. Thus the exponential has no maximum or minimum, the hyperbolic cosine has a minimum but no maximum, and the function $x \mapsto \arctan(x) + \sin(x)$ has infinitely many local maxima and minima, but no maximum or minimum.

By a **neighbourhood of infinity** we mean a ray $\{x \in \mathbb{R} : x > a\}$, where a is a real number. A neighbourhood of $-\infty$ is defined similarly, and both points at infinity $\pm\infty$ are handled at once with the construct $\{x \in \mathbb{R} : |x| > a\}$. The points $\pm\infty$ are not numbers but they have neighbourhoods, which make them more tangible. So the sentence

The function f is constant in a neighbourhood of infinity.

means that f is constant for all sufficiently large values of the argument. It may be written symbolically as

$$\exists a \in \mathbb{R}, \quad \forall x \in \mathbb{R}^+, \quad f(a+x) = f(a)$$

or as

$$\exists a \in \mathbb{R}, \quad \forall x > a, \quad f(x) = f(a).$$

5.4.1 Neighbourhoods and Sets

Neighbourhoods are instrumental to the description of sets of numbers. This is an appealing part of the mathematical dictionary, due to the vivid mental pictures we associate with a geometric language. The case for expanding our dictionary is easily made:

$$X_1 = \{1/n : n \in \mathbb{N}\} \quad X_2 = \mathbb{Q} \cap (0,1).$$

How can we describe such sets?

Let $X \subset \mathbb{R}$. A point $x \in X$ is **isolated** if there is a neighbourhood of x that contains no other point of X. The set $\mathbb{Z}$ consists entirely of isolated points, and so does the set X_1 above. By contrast, X_2 has no isolated points.

A point x is an **interior point** of a set X if X contains a neighbourhood of x. A point x is a **boundary point** of a set X if every neighbourhood of x contains points of X as well as points of the complement of X. An isolated point is necessarily a boundary point, but not all boundary points are isolated. The boundary points of an interval are its end-points; all other points are interior points and there are no isolated points. Neither X_1 nor X_2 have interior points; the origin is a limit point of X_1.

A set is **closed** if it contains all its boundary points, and is **open** if all its points are interior points. For intervals, these concepts agree with those given in Sect. 2.1.3.

5.4 Neighbourhoods

The sets X_1 and X_2 are neither open nor closed. The **closure** of a set X, denoted by $\overline{X}$, is the union of X and the boundary points of X. We see that $\overline{X}_1 = X_1 \cup \{0\}$ and $\overline{X}_2 = [0, 1]$.

5.5 Continuity

Continuity is a fundamental concept in the theory of functions. Loosely speaking, a function is continuous if it has 'no jumps', but this naive definition is only adequate for simple situations. We begin by considering continuity at a specific point of the domain of a function. An informal—yet accurate—characterisation of continuity is to say that a function is continuous at a point if its value there can be inferred unequivocally from the values at neighbouring points. Thus neighbourhoods come into play.

For example, the value of the sign function (5.3) at the origin cannot be inferred from the surrounding environment; even if we defined, say, $\mathrm{sign}(0) = 1$, the ambiguity would remain. Things can get a lot worse. Consider the real function

$$x \mapsto \begin{cases} 1 & \text{if } x \in \mathbb{Q} \\ 0 & \text{if } x \notin \mathbb{Q}. \end{cases} \tag{5.9}$$

Given that in any neighbourhood of any real number there are both rational and irrational numbers, there is no way of inferring the value of this function at any point by considering how the function behaves in the surrounding region.

To write a symbolic definition of continuity we need to quantify neighbourhoods. For this purpose, let us denote by $\mathcal{N}_x$ the set of all neighbourhoods of a point x in $\mathbb{R}$. (This is an **abstract set**, which may be represented as a set of end-points of real intervals—see Sect. 2.3.3 and Exercise 5.12.)

We say that f is **continuous at the point** a if for every neighbourhood J of $f(a)$ we can find a neighbourhood of a whose image is contained in J. In symbols:

$$\forall J \in \mathcal{N}_{f(a)}, \ \exists I \in \mathcal{N}_a, \ f(I) \subset J. \tag{5.10}$$

This elaborate expression may be written in short-hand notation as

$$\lim_{x \to a} f(x) = f(a)$$

which says that $f(a)$ is the **limit** of $f(x)$ as x **tends to** a.

Let's go through Definition (5.10) in slow motion. The set $\mathcal{N}_a$ consists of all open intervals containing the point a where continuity is being tested. The set $\mathcal{N}_{f(a)}$ consists of all neighbourhoods of $f(a)$, the value of the function at a. An arbitrary interval J containing $f(a)$ is given, typically very small. We must now find an interval I containing a whose image fits inside J. The choice of I will depend on J, and if we succeed in all cases, then the function is continuous at a. We don't need to know what the set $f(I)$ looks like; it suffices to know that $f(I)$ can be made small by making I small.

A function that is not continuous at a point of the domain is said to be **discontinuous**. A function is **discontinuous everywhere** if it has no points of continuity, like the function (5.9).

Continuity may be defined without reference to neighbourhoods, but in this case one must supply quantitative information. The full notation is considerably more involved than (5.10):

$$\forall \varepsilon \in \mathbb{R}^+, \ \exists \delta \in \mathbb{R}^+, \ \forall x \in \mathbb{R}, \ |x - a| < \delta \Rightarrow |f(x) - f(a)| < \varepsilon$$

and it remains so even if we strip it of all references to $\mathbb{R}$:

$$\forall \varepsilon > 0, \ \exists \delta > 0, \ \forall x, \ |x - a| < \delta \Rightarrow |f(x) - f(a)| < \varepsilon. \qquad (5.11)$$

A function is **continuous** if it is continuous at all points of the domain. An additional quantifier is required in Definition (5.10):

$$\forall a \in A, \ \forall J \in \mathcal{N}_{f(a)}, \ \exists I \in \mathcal{N}_a, \ f(I) \subset J, \qquad (5.12)$$

where A is the domain of f.

A function f is **differentiable at** a if the limit

$$\lim_{x \to a} F(x) \quad \text{where} \quad F(x) = \frac{f(x) - f(a)}{x - a} \quad x \neq a$$

exists. The function F is called the **incremental ratio** of f at a, which is not defined at $x = a$. However, if f is differentiable at a, then by letting $F(a) \stackrel{\text{def}}{=} \lim_{x \to a} F(x)$, the function F becomes **continuous** at a. Let's sum this up in a sentence.

> *A function is differentiable at a point if its incremental ratio is continuous at that point; in this case the value of the incremental ratio is the derivative of the function.*

A function is said to be **differentiable** if it is differentiable at all points of its domain. A function that is differentiable **sufficiently often** (all derivatives up to a sufficiently high order exist) is said to be **smooth**. The expression 'sufficiently often' is deliberately vague; its precise meaning will depend on the context.

Many elementary real functions are continuous in their respective domains: the polynomials, the trigonometric functions, and the exponential and logarithmic functions, etc. These functions are also differentiable infinitely often.

A function f is **singular** at a point x if f is undefined at x, in which case x is a **singularity** (or a **singular point**) of the function. So the function $x \mapsto 1/x$ is singular at the origin. A function that is not singular is said to be **well-behaved** or **regular**, the latter term used mostly for functions of a complex variable. We write:

5.5 Continuity

A rational function has finitely many singularities: the roots of the polynomial at the denominator. This function is regular everywhere else.

The tangent, secant, and co-secant have infinitely many singularities, evenly spaced along the real line.

In analysis the term 'singular' is also used for well-defined functions that exhibit certain esoteric pathologies.

5.6 Other Properties

A function of the form $f(x) = ax$, where a is a real number, is said to be **linear**. The term linear originates from 'line', and functions of the type $f(x) = ax + b$, are sometimes referred to as being 'linear' because their graph is a line (the correct term is **affine**). This acceptation of the term linear is common for polynomial functions, which in the first instance are characterised by their degree. So we speak of a **linear**, **quadratic**, **cubic**, **quartic** function, etc.

Consider the absolute value function, defined as follows:

$$|x| = \begin{cases} x & \text{if } x \geq 0 \\ -x & \text{if } x < 0. \end{cases} \tag{5.13}$$

This function is not linear, but it is made of two linear pieces, glued together at the origin. Functions made of linear or affine pieces are said to be **piecewise linear** or **piecewise affine** [see Eq. (5.3) and Fig. 5.3]. A function is **piecewise defined** if its domain is partitioned into disjoint intervals or rays, with the function being specified independently over each interval. The properties of a piecewise-defined function may fail at the end-points of the intervals of definition. In this case terms such as **piecewise increasing**, **piecewise continuous**, **piecewise differentiable**, **piecewise**

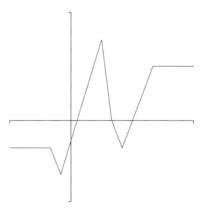

Fig. 5.3 A continuous piecewise affine function

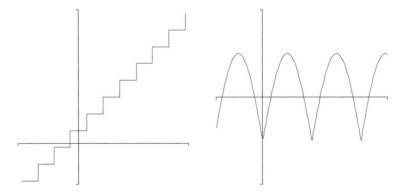

Fig. 5.4 *Left* A step function. (The *vertical* segments are just a guide to the eye; they are not part of the graph of the function.) *Right* A periodic continuous function which is piecewise differentiable

smooth may be used—see Fig. 5.4. A piecewise constant function is called a **step-function**.

EXAMPLE. The **floor** of a real number x, denoted by $\lfloor x \rfloor$, is the largest integer not exceeding x. Similarly, the **ceiling** $\lceil x \rceil$ represents the smallest integer not smaller than x. Floor and ceiling are the prototype step functions. Closely connected to the floor is the **fractional part** of a real number x, denoted by $\{x\}$. (The notational clash with the set having x as its only element is one of the most spectacular in mathematics!) The fractional part is defined as

$$\{x\} := x - \lfloor x \rfloor$$

from which it follows that $0 \leqslant \{x\} < 1$. This function is discontinuous and piecewise affine.

EXAMPLE. Describe the following function: [≰]

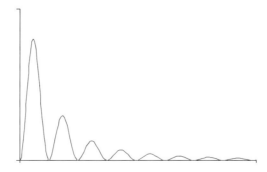

5.6 Other Properties

This is a smooth function, which is bounded and non-negative. It features an infinite sequence of evenly spaced local maxima, whose height decreases monotonically to zero for large arguments. There is one zero of the function between any two consecutive local maxima.

We rewrite it, borrowing some terminology from physics.

This function displays regular oscillations of constant period, with amplitude decreasing monotonically to zero.

The following functions have a behaviour that is qualitatively similar to that displayed above.

$$f(x) = \sin^2(x) e^{-x} \qquad f(x) = \frac{1 - \cos(x)}{1 + x^2}.$$

5.7 Describing Sequences

A sequence may be thought of as a function defined over the natural numbers, or, more generally, over a subset of the integers (Sect. 3.1). Indeed in number theory the real (or complex) sequences are called **arithmetical functions**. Using this analogy, some terminology introduced for real functions translates literally to real sequences. So the terms

positive, negative, increasing, decreasing, monotonic, constant, periodic, bounded, bounded away from zero

have the same meaning for sequences as they have for functions. We write

The sequence of primes is positive, increasing, and unbounded.

Other terms require amendments or are simply not relevant to sequences. For instance, injectivity is not used—we just say that the terms of a sequence are distinct—while surjectivity is rarely significant. Invertibility is used in a different sense than for functions—see Sect. 9.4. The terms 'even' and 'odd' are still applicable to doubly-infinite sequences.

A sequence which settles down to a periodic pattern from a certain point on is said to be **eventually periodic**. More precisely, a sequence $(x_1, x_2, \ldots)$ is eventually periodic if for some $k \in \mathbb{N}$ the sub-sequence $(x_k, x_{k+1}, \ldots)$ is periodic. If the periodic pattern consists of a single term, then the sequence is said to be **eventually constant**.

The sequence of digits of a rational number is eventually periodic.

The idea of continuity does not apply to sequences because the concept of neighbourhood of an integer is pointless. Except at infinity, that is. The **neighbourhoods of infinity** are the infinite sets of the form $\{k \in \mathbb{N} : k > K\}$, for some integer K, and we use the expression **for all sufficiently large** k, to mean for all $k > K$, for some K.

We can now test continuity at infinity. Adapting Definition (5.10) to the present situation, we say that the sequence (a_k) **converges** to the **limit** c if, given any neighbourhood J of c, all terms of the sequence eventually belong to J. In symbols:

$$\forall J \in \mathcal{N}_c, \ \exists K \in \mathbb{N}, \ \forall k > K, \ a_k \in J.$$

This expression is written concisely as

$$\lim_{k \to \infty} a_k = c.$$

By stipulating that $a_\infty = c$, we see that a real sequence **converges** if it's **continuous at infinity**. In the special case $c = \infty$ we say that the sequence **diverges**.

Exercise 5.1 Write each symbolic sentence in two ways:
(i) without any symbol, apart from f.
(ii) with symbols only, using quantifiers. (You may assume that $f \colon \mathbb{R} \to \mathbb{R}$.)

1. $f(0) \in \mathbb{Q}$
2. $f(\mathbb{R}) = \mathbb{R}$
3. $\# f(\mathbb{R}) = 1$
4. $f(\mathbb{Z}) = \{0\}$
5. $0 \in f(\mathbb{Z})$
6. $f^{-1}(\{0\}) = \mathbb{Z}$
7. $f(\mathbb{R}) \subset \mathbb{Q}$
8. $f(\mathbb{R}) \supset \mathbb{Z}$
9. $f(\mathbb{Z}) = f(\mathbb{N})$
10. $f(\mathbb{Q}) \cap \mathbb{Q} = \emptyset$
11. $f^{-1}(\mathbb{Q}) = \emptyset$
12. $\# f^{-1}(\mathbb{Z}) < \infty$.

Exercise 5.2 For each symbolic sentence of the previous exercise, give examples of functions for which the sentence is true and functions for which the sentence is false.

Exercise 5.3 Write each symbolic sentence without symbols, apart from f.
1. $\forall x \in \mathbb{Z}, \ f(2x) = 0$
2. $\forall x \in \mathbb{R}, \ f(f(x)) = x$
3. $\forall x \in \mathbb{R}^+, \ f(-x) = 0$
4. $\forall x \in [0, 1], \ f(x) \neq 0$
5. $\forall x \in \mathbb{Q} \setminus \{0\}, \ f(x) \neq 0$
6. $\forall x \in \mathbb{N}, \ f(x) \notin \mathbb{Q}$
7. $\forall x \in \mathbb{Z}, \ \exists y \in \mathbb{N}, \ f(x + y) = 0$
8. $\forall x \in \mathbb{R}, \ \exists y \in \mathbb{R}, \ |f(y)| < |f(x)|/2$
9. $\forall x \in \mathbb{R}, \ f(x) = 4 f(x/2)$.

5.7 Describing Sequences

Exercise 5.4 Turn words into symbols.

1. The function f has no zero.
2. The function f is not always positive.
3. The function f does not vanish except, possibly, at the origin.
4. The function f vanishes identically outside the open unit interval.
5. The zeros of f include all multiples of π.
6. The function f may vanish only at some rational.
7. There are zeros of f arbitrarily close to the origin.
8. The zeros of f are isolated.
9. There is a zero of f between any two consecutive integers.
10. The function f has no zero for all sufficiently large arguments.
11. Every integer is a local maximum of f.
12. The function f is locally invertible near the origin.

Exercise 5.5 Consider the following implications, where f is a real function.

1. If f is decreasing, then $-f$ is increasing.
2. If f is decreasing, then $|f|$ is increasing.
3. If $|f|$ is increasing, then f is monotonic.
4. If f is even, then $|f|$ is also even.
5. If f is unbounded, then f is surjective.
6. If f is continuous, then $|f|$ is continuous.
7. If f is differentiable, then $|f|$ is differentiable.
8. If f is periodic, then f^2 is periodic.

Of each implication:

(a) state the converse, and decide whether it's true or false;
(b) state the contrapositive, and decide whether it's true or false;
(c) state the negation.

Exercise 5.6 As in Exercise 4.3.

1. Every differentiable real function is continuous.
2. A bounded real function cannot be injective.
3. The sum of two odd functions is an odd function.
4. No polynomial function is bounded.
5. The composition of two periodic functions is periodic.
6. Any function with an increasing derivative is increasing.

Exercise 5.7 Consider the Definition (5.5) of a periodic function. What happens if we place the quantifiers in reverse order?

$$\forall x \in \mathbb{R}, \ \exists T \in \mathbb{R}^*, \ f(x+T) = f(x) \qquad (5.14)$$

$$\forall x \in \mathbb{R}, \ \exists T \in \mathbb{R}^+, \ f(x+T) = f(x). \qquad (5.15)$$

1. Find a function that satisfies (5.14) but not (5.5).
2. Find a function that satisfies (5.15) but not (5.5).
3. Characterize the set of functions that satisfy (5.14). Do the same for (5.15).

Exercise 5.8 Describe the behaviour of the following functions. [≴, 30]

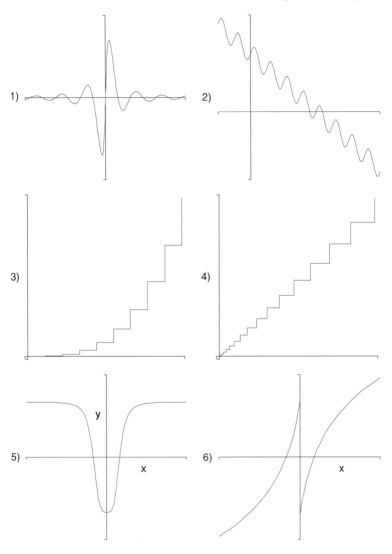

5.7 Describing Sequences

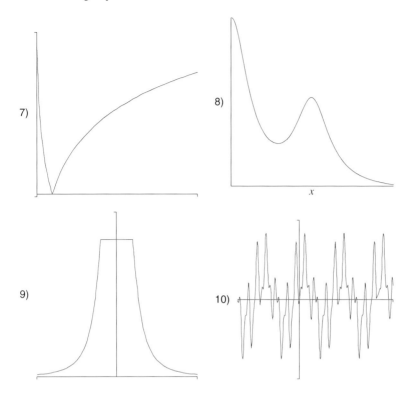

Exercise 5.9 By experimenting on a computer, if necessary, define real functions whose behaviour is qualitatively similar to those of the previous problem.

Exercise 5.10 Give an example of a real function with the stated properties.

1. Increasing, with decreasing derivative.
2. Unbounded, with unbounded derivative.
3. Unbounded, with bounded derivative.
4. Discontinuous, with continuous absolute value.
5. With infinitely many zeros, but not periodic.
6. Unbounded and vanishing infinitely often.
7. The image is a half-open interval.
8. The image is the set of natural numbers.

Exercise 5.11 Express each statement with symbols.

1. The sequences (a_k) and (b_k) are distinct.
2. The sequence (a_k) is eventually constant.
3. The sequence (a_k) is not periodic.
4. The sequence (a_k) is eventually periodic.
5. The sequence (a_k) has infinitely many negative terms.
6. Eventually, all terms of the sequence (a_k) become negative.

7. *The terms of the sequence (a_k) get arbitrarily close to zero.*
8. *The sequence (a_k) does not converge.*
9. *Each term of the sequence (a_k) appears at least twice.*
10. *Each term of the sequence (a_k) appears infinitely often.*
11. *Each natural number appears infinitely often in the sequence (a_k).*
12. *The sequence (a_k) has a bounded sub-sequence.*
13. *The sequence (a_k) has a convergent sub-sequence.*

Exercise 5.12 Find a representation for the abstract set $\mathcal{N}_x$ of all neighbourhoods of a point $x \in \mathbb{R}$ (see Sect. 2.3.3).

Chapter 6
Writing Well

In this chapter we consider some techniques for writing mathematics. We deal with small-scale features: choosing an appropriate terminology and notation, writing clear formulae, mixing words and symbols, writing definitions, introducing a new concept. These are pre-requisites for the more substantial task of structuring and delivering a mathematical argument, to be tackled in later chapters.

A universal problem besets any form of specialised writing: how much explanation should we provide? Ideally, we ought to explain the meaning of every word that we use. But this is impossible: we would need to explain the words used in the explanation, and so on. Instead, we should only explain a word or symbol if our explanation will make it clearer than it was before. Accordingly, we shall call a word or symbol **primitive**[1] if it's suitable to use without an explanation of its meaning.

An ordinary English word like 'thousand' is obviously primitive, but for more specialised words we must consider the context. When we communicate to the general public, the term **multiplication** can safely be regarded as primitive. Likewise, there should be no need to explain to a mathematician what an **eigenvalue** is, while a number theorist will be familiar with **conductor**. Then there are extremes of specialisation: only a handful of people on this planet will know the meaning of **Hsia kernel**. Finally, terms such as **exceptional set** mean different things in different contexts. Understanding what constitutes an appropriate set of primitives is an essential pre-requisite for the communication of complex knowledge. Getting it right is tricky.

The writing developed in this chapter is addressed to a mathematically mature audience. Occasionally, we will consider writing for the general public.

6.1 Choosing Words

Precision must be the primary concern of anyone who writes mathematics. We list some common mistakes and inaccuracies that result from a poor choice of words.

BAD: the equation $x - 3 \leqslant 0$
GOOD: the inequality $x - 3 \leqslant 0$

---

[1] This terminology is due to Blaise Pascal (French: 1623–1662).

BAD: the equation $x^2 - 1 = (x - 1)(x + 1)$
GOOD: the identity $x^2 - 1 = (x - 1)(x + 1)$
BAD: the identity $x = \sqrt{x^2}$
GOOD: the equation $x = \sqrt{x^2}$
BAD: the solution of $x^k = x$
GOOD: a solution of $x^k = x$
BAD: the minima of a quadratic function
GOOD: the minimum of a quadratic function
BAD: the function $\sin(x)$
GOOD: the sine function
BAD: the function $f(A)$ of the set A
GOOD: the image of the set A under f
BAD: the interval $[1, \infty)$
GOOD: the ray $[1, \infty)$ (the infinite interval $[1, \infty)$)
BAD: the set $\mathbb{Z}$ minus $k\mathbb{Z}$
GOOD: the set difference of $\mathbb{Z}$ and $k\mathbb{Z}$
BAD: the area of the unit circle
GOOD: the area of the unit disc
BAD: the coordinates of a complex number
GOOD: the real and imaginary parts of a complex number
BAD: the absolute value is positive
GOOD: the absolute value is non-negative
BAD: the function crosses the vertical axis at a positive point
GOOD: the graph of the function intersects the ordinate axis at a positive point.

Once our writing is accurate, we consider refining the choice of words to differentiate meaning, improve legibility, or avoid repetition.

A set is mostly called a set, although **collection** or **family** are useful alternatives (as in the title of Sect. 2.3.1). A set becomes a **space** if it has an added structure, like a **metric space** (a set with a distance) or a **vector space** (a set with an addition and a scalar multiplication).

The word **element** expresses a special kind of subsidiary relationship. If A is a set and $a \in A$, then we say that a is an element of A. The term **member** is a variant, and so is **point**, which is the default choice for geometrical sets. So if $A \subset \mathbb{R}^n$, then x is a **point** of A. For sequences, we may replace 'element' by **term**. This is necessary if the elements of a sequence are added (multiplied) together, in which case the operands are the terms of the sum (product), not the elements. However, if $V = (v_k)$ is a vector, then v_k is a **component** of V, not a term (even though a vector is just a finite sequence). But if $M = (m_{i,j})$ is a matrix, then $m_{i,j}$ is an element or an **entry** of M, not a component (even though a matrix is just a sequence of vectors).

The word **variable** is used in connection with functions and equations. In the former case it has the same meaning as **argument**, and it refers to the function's input data. In the latter case it means **unknown**—a quantity whose value is to be found. Polynomials and rational functions may represent functions or algebraic objects; for the latter, the term **indeterminate** is preferable to variable.

6.1 Choosing Words

The term **parameter** is used to identify a variable which is assigned a value that remains fixed in the subsequent discussion. For example, the indefinite integral of a function is a function which depends on a parameter, the integration constant:

$$\int g(x)dx = f(x) + c.$$

The two symbols x and c play very different roles here, so we have a **one-parameter family** of functions, rather than a function of two variables.

A function is not always called a function. This may be true for real functions, but in more general settings the terms **map**, **mapping**, and **transformation** are at least as common. So a complex function $\mathbb{C} \to \mathbb{C}$ may be called a mapping, and a function between euclidean spaces $\mathbb{R}^n \to \mathbb{R}^m$ is a mapping or a transformation. Some functions are called **operators**. These include basic functions, like addition of numbers or intersection of sets, represented by specialised symbols ($+, \cap$) and syntax ($x + y$ rather than $+(x, y)$). The **arithmetical operators**, the **set operators**, the **relational operators**, and the **logical operators** are functions of this type. The attribute **binary** means that there are two operands, while **unary** is used for a single operand, like changing the sign of a number. The term **operator** is also used to characterise functions acting on functions which produce other functions as a result. The **differentiation operator** is a familiar example. A real-valued function acting on functions is called a **functional**. So definite integration is a functional.

In logic, a function is called a **predicate**, a **boolean function**, or a **characteristic function**. These terms represent the same thing, but they are not completely interchangeable. 'Characteristic function' should be used if there is explicit reference to the associated set, and 'predicate' (or 'boolean function') otherwise. If the arguments of a function are also boolean, then there is a strong case for using 'boolean function'. Some examples:

> The negation of a predicate is a predicate.
> Let χ be the characteristic function of the prime numbers.
> Let us consider the boolean function $(x, y) \mapsto \neg x \wedge y$.
> The predicate '$n \mapsto (3 \mid n(n + 1))$' is the characteristic function of a proper subset of the integers.
> A relational operator is a boolean function of two variables.

In mathematics it is acceptable to change the meaning of established words. The following re-definitions of universal geometrical terms were taken from mathematics research literature:

> By a **triangle** we mean a metric space of cardinality three.
> By a **segment** we mean a maximal subpath of P that contains only light or only heavy edges.
> By a **circle** we mean an affinoid isomorphic to $\max \mathbf{C}_p(T, T - 1)$.

Re-defining such common words requires some self-confidence, but in an appropriate context this provocative device may be quite effective. Clearly the new meaning is meant to remain confined to the document in which it's introduced.

6.2 Choosing Symbols

Choosing mathematical notation is difficult. Mathematicians are notoriously reluctant to accept standardisation of notation, to a degree unknown in other disciplines. Indeed, the ability to adjust quickly to new notation is regarded as one of the skills of the trade. The reality is somewhat different: absorbing new notation requires effort, and most people would gladly avoid it. So if we intend to communicate mathematics without confusing or alienating our audience, the notation must be simple, logical, and consistent.

Two golden rules should inform the use of symbols:

- DO NOT INTRODUCE UNNECESSARY SYMBOLS.
- DEFINE EVERY SYMBOL BEFORE USING IT.

Once defined, a symbol should be used consistently. Never use the same symbol for two different things or two symbols for the same thing, even in instances appearing far apart in the document. Don't write 'A_j, for $1 \leqslant j < n$' in one place and 'A_k, for $1 \leqslant k < n$' in another, unless there is a good reason to do so. Such small inconsistencies produce some 'notational pollution'. As the pollution piles up, reading becomes tiresome.

In this section we offer guidelines on how to choose symbols. These are not laws, and may be adapted to one's taste or rejected. What's important is to develop awareness of notation, and to make conscious decisions about it.

SETS. Represent sets by capital letters, Roman or Greek, such as

$$S \quad \mathscr{A} \quad \Omega.$$

The large variety of fonts available in modern typesetting systems increases our choice. When dealing with generic sets, then A, B, C or X, Y, Z are good symbols. For specific sets, choose a symbol that will remind the reader of the nature of the set. So, for an **alphabet** $\{a, b, c, \ldots\}$, the symbols $A, \mathscr{A}$ are obvious choices; likewise, F is appropriate for a set of functions, etc.

Lower-case symbols like x, y represent the elements of a set. So $x \in A$ is a good notation, $X \in A$ is bad, and $X \in a$ is very bad. If more than one set is involved, consider using matching symbols. Thus

$$a \in A \quad b \in B \quad c \in C$$

is more coherent than

$$x \in A \quad y \in B \quad z \in C.$$

6.2 Choosing Symbols

Some thought may be required for sets of sets. Consider the expressions

$$f(A \cap B) \qquad f(x \cap y).$$

The left expression will be interpreted as the image of the intersection of two sets under a function. In this case $A \cap B$ represents a subset of the domain of f. However, suppose that the domain of f is a set of sets (e.g., a power set, see Sect. 2.3). Then the argument of f is a set, and this expression becomes dangerously ambiguous. The notation in the right expression removes this ambiguity. The combination of standard symbols x, y for variables and a set operator makes it clear that the argument $x \cap y$ of the function is an element, rather than a subset, of the domain.

INTEGERS. When choosing a symbol for an integer, begin from the middle region of the Roman alphabet

$$i, j, k, l, m, n \qquad (6.1)$$

particularly if an integer is used as subscript or superscript.[2] However, use p for a prime number, and q if there is a prime different from p. The list of adjacent letters (6.1) cannot be extended; the preceding symbols, f, g, h, are typical function names (see below), while the symbol that follows, o, is rarely used, not only for its resemblance to 0 (zero), but also because it has an established meaning in asymptotic analysis (it appears in expressions of the form $o(\log x)$). A capital letter in the list (6.1) may be used to represent a large integer, or combined with lower-case letters to denote an integer range: $n = 1, \ldots, N$. This combination of symbols is much used in connection with sums and products (Sect. 3.2).

RATIONALS. For rational numbers, use lower-case Roman letters in the ranges a–e, or p–z. The notation

$$r = \frac{m}{n}$$

is good, because 'r' reminds us of 'rational', while numerator and denominator conform to the convention for integers. If there is more than one rational, use adjacent symbols, s, t in this case.

REAL NUMBERS. For real numbers, use the same part of the Roman alphabet as for rationals, or the Greek alphabet:

$$\alpha, \beta, \gamma, \ldots$$

If there are both rational and real numbers and if the distinction between them is important, then use Roman for the rationals and Greek for the reals.

Some Greek symbols have preferential meaning: small quantities are usually represented by ε, δ, while for angles one uses ϕ, φ, θ. Famous real constants have dedicated symbols:

---

[2] Physicists use Greek letters for subscripts and superscripts.

$$\pi = 3.141592653\ldots \quad \textbf{Archimedes' constant (or Pi)}$$
$$e = 2.718281828\ldots \quad \textbf{Napier's constant}^3 \quad\quad (6.2)$$
$$\gamma = 0.333177924\ldots \quad \textbf{Euler-Mascheroni constant.}^4$$

These constants are typeset with the 'upright' typeface, to highlight their distinguished status. Upright fonts are known as **roman** (even for Greek letters), as opposed to **italic**, which are slanted fonts: π, e, γ.

COMPLEX NUMBERS. Complex numbers tend to occupy the end of the Roman alphabet, and your first choice should be z or w. On the complex plane, we write $z = x + \mathrm{i} y$, where x and y are the real and imaginary parts of z, and i is the **imaginary unit**, again typeset in roman font. (However, number theorists use $\sqrt{-1}$, not i.) The standard notation for a complex number in polar coordinates is $z = \rho \mathrm{e}^{\mathrm{i}\theta}$, which also combines roman and italic fonts.

UNKNOWNS. The quintessential symbol for an equation's unknown is x, invariably followed by y and z if there are more unknowns. For a large number of unknowns it is necessary to use sequence notation $x_1, \ldots, x_n$. This notation is also appropriate for the indeterminates of a polynomial and for the arguments of a function.

COMPOSITE OBJECTS. Objects which have constituent elements (groups, graphs, matrices) are best represented with capital letters, Roman or Greek. So use G or Γ for a group or a graph, and M for a matrix. If you have two groups, use adjacent symbols, like G and H. As with sets, for these objects' components consider using matching symbols, e.g., $g \in G$. A notable exception are graphs, where v and e are invariably used for vertices and edges, respectively.

FUNCTIONS. The default choice for a function's name is, of course, f, and if there is more than one function, use the adjacent symbols g, h. Lower-case symbols work well with any number of variables: $f(x)$, $f(x, y, z)$, $f(x_1, \ldots, x_n)$. But if the co-domain of a function is a cartesian product, then the function is a vector, and capital letters are preferable. So a real function of two variables may be specified as

$$F : \mathbb{R}^2 \to \mathbb{R}^2 \quad\quad (x, y) \mapsto (f_1(x, y), f_2(x, y)).$$

Greek symbols, either capital or lower-case, are also commonly used for functions' names. The contrast between Roman and Greek symbols may be exploited to separate out the symbols' roles, as in $\mu(x)$ or $f(\lambda)$.

Some famous functions are named after, and represented by, a symbol (often a Greek one), thereby creating a strong bond between object and notation. The best known are **Euler's gamma function Γ**

$$\Gamma(s) = \int_0^\infty x^{s-1} \mathrm{e}^{-x} \mathrm{d}x$$

---

[3] John Napier (Scott: 1550–1617).
[4] Lorenzo Mascheroni (Italian: 1750–1800).

6.2 Choosing Symbols

(the extension of the factorial function to complex arguments), and **Riemann's zeta-function**[5] ζ

$$\zeta(s) = \sum_{n=1}^{\infty} \frac{1}{n^s}. \tag{6.3}$$

There is a peculiar notation for this function: its complex argument s is commonly written as $s = \sigma + i\tau$, with σ and τ real numbers. Other functions with dedicated notation are Euler's φ-function (Eq. 3.8), Dedekind's η-function, Kroneker's δ-function, Weierstrass' $\mathscr{P}$-function, Lambert's $\mathscr{W}$-function, etc.

SEQUENCES AND VECTORS. Sequences pose specific notational problems due to the presence of indices. Consider the various possibilities listed in (3.2): which one should we choose? Be guided by the principle of economy: a symbol should be introduced only if it's strictly necessary. So the notation (a_k) is quite adequate for a generic sequence, or if the specific properties of the sequence (the initial value of the index, its finiteness) are not relevant. When more information is needed, the notation $(a_k)_{k \geq 1}$ is more economical than $(a_k)_{k=1}^{\infty}$, but the latter may be a better choice if it is to be contrasted with $(a_k)_{k=1}^{n}$. In turn, the latter is not as friendly as $(a_1, \ldots, a_n)$, although it is more concise.

If a sequence is referred to often, even the stripped down notation (a_k) could become heavy, and it may be advisable to allocate a symbol for the sequence:

$$a = (a_1, a_2, \ldots) \qquad \mathbf{v} = (v_1, \ldots, v_n).$$

As usual, we have employed matching symbols, using, respectively, a minimalist lower-case Roman character and a lower-case boldface character which is common for vectors. When using ellipses, two or three terms of the sequence usually suffice, but there are circumstances where more terms or a different arrangement of terms is needed.

For example, in the expression

$$(1 + x, 1 + x^2, \ldots, 1 + x^{2^k}, \ldots)$$

the insertion of the general term removes any ambiguity, while the ellipsis on the right suggests that the sequence is infinite—cf. (3.1). The notation

$$(a_1, \ldots, a_{k-1}, a_{k+1}, \ldots, a_n)$$

denotes a sub-sequence of a finite sequence obtained by deleting the k-th term, for an unspecified value of $k \neq 1, n$.

Things get complicated with sequences of sequences. This situation is not at all unusual; for instance, we may have a sequence of vectors whose components must be referred to explicitly. We write

---

[5] Bernhard Riemann (German: 1826–1866).

$$V = (V_1, V_2, \ldots) \quad \text{or} \quad \mathbf{v} = (\mathbf{v}_1, \mathbf{v}_2, \ldots).$$

Let V_k (or $\mathbf{v}_k$) be the general term of our sequence. How are we to represent its components? As usual, we choose the matching symbol v, with a subscript indicating the component. However, the integer k has to appear somewhere, and its range must be specified. It is advisable to keep k out of the way as much as possible:

$$V_k = (v_1^{(k)}, \ldots, v_n^{(k)}) \quad k = 1, 2, \ldots.$$

In this expression we have used ellipses to specify the ranges of the indices; we could have used inequalities as well:

$$V_k = (v_j^{(k)}) \quad 1 \leqslant j \leqslant n, \quad k \geqslant 1.$$

The parentheses are obviously needed for the superscript, for otherwise v_i^k would be interpreted as v_i raised to the k-th power. However, it may just happen that we need to raise the vector components to some power. Clearly we can't use adjacent superscripts $v_3^{(2)\,4}$, so parentheses are needed, but the straightforward notation $(v_3^{(2)})^4$ is awkward. For a more elegant solution, we represent k as an additional subscript, adopting, in effect, matrix notation:

$$V_k = (v_{1,k}, \ldots, v_{n,k}) \quad k \geqslant 1.$$

As a side note, one should keep in mind that with vectors the multiplication symbols '·' and '×' are reserved for the scalar and vector products, respectively—cf. (2.10). Hence for scalar multiplication we must use juxtaposition:

$$a(bV \cdot cW) \qquad x\mathbf{v} \times y\mathbf{u}.$$

DERIVED SYMBOLS. Closely related objects require closely related notation. Proximity in the alphabet, e.g., x, y, z may be used for this purpose. For a stronger bond, the meaning of a symbol may be modified using subscripts, superscripts and other decorations:

$$A^* \quad \bar{\eta} \quad n^+ \quad \underline{h} \quad \tilde{e} \quad \Omega_- \quad Z_r.$$

The derived symbol $\mathbb{N}_0 = \mathbb{N} \cup \{0\}$ is often found in the literature. Many symbols derived from $\mathbb{R}$ are in use:

$$\mathbb{R}^+ = \{x \in \mathbb{R} : x > 0\} \qquad \mathbb{R}_{\geqslant 0} = \{x \in \mathbb{R} : x \geqslant 0\}$$
$$\mathbb{R}^* = \mathbb{R} \setminus \{0\} \qquad \overline{\mathbb{R}} = \mathbb{R} \cup \{\infty\}.$$

6.2 Choosing Symbols

These sentences illustrate the use of derived symbols:

> Let $f : X \to X$ be a function, and let $x^* = f(x^*)$ be a fixed point of f.
> We consider the endpoints x_- and x_+ of an interval containing x.
> Let f be a real function, and let

$$f^+ : \mathbb{R} \to \mathbb{R} \qquad x \mapsto \begin{cases} f(x) & \text{if } f(x) \geq 0 \\ 0 & \text{if } f(x) < 0. \end{cases}$$

It must be noted that there is no general agreement on the meaning of decorations. Thus for a set the over-bar denotes the so-called **closure**—adjoining to a set all its boundary points, see Sect. 5.4.1 (the transformation from $\mathbb{R}$ to $\overline{\mathbb{R}}$ is essentially a closure operation). However, for complex numbers the over-bar denotes complex conjugation. If f is a function, then f' is the derivative of f, but for sets a prime indicates taking the complement.

EXAMPLE. We illustrate notational problems raised by the coexistence of variables and parameters. For a fixed value of z, the bivariate polynomial

$$f(x, z) = -z^2 + xz + 1$$

becomes a polynomial in x, and we wish to adopt a notation that reflects the different roles played by the symbols x and z. We replace z with a, to keep it far apart from x in the alphabet, and we rewrite the expression above as

$$f_a(x) = ax + 1 - a^2 \qquad a \in \mathbb{R}. \tag{6.4}$$

We now have a **one-parameter family** of linear polynomials in x. For fixed a, the equation $y = f_a(x)$ is the cartesian equation of a line, so we also have a one-parameter family of lines in the plane. Plotting some of these lines reveals a hidden structure: they form the **envelope** of a parabola (Fig. 6.1). Likewise, if we fix $x = a$ we obtain a one-parameter family of quadratic polynomials $g_a(z) = -z^2 + az + 1$.

6.3 Improving Formulae

When the physicist Stephen Hawking was writing his book *A Brief History of Time*, an editor warned him that every new equation in the book would halve the readership. So he included a single equation. When writing for the general public, one should expect diffidence—even hostility—towards formulae. Even though mathematicians are trained to deal with formulae, it is still safe to assume that no one likes to struggle with too many symbols. In this section we explore techniques to improve the clarity of formulae, through presentation, notation, and layout.

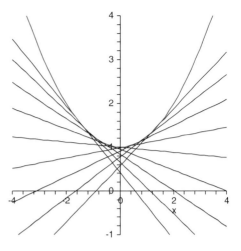

Fig. 6.1 The one-parameter family of lines $y = f_a(x)$, where f_a is given by (6.4). These lines are tangent to the parabola $y = x^2/4 + 1$

Formulae may be embedded in the text or displayed, and in a document one usually finds both arrangements. In either case a formula must obey standard punctuation rules. The following passage features embedded formulae with appropriate punctuation.

> For each $x \in X$ we have the decomposition $x = \xi + \lambda$, with $\xi \in \Xi$ and $\lambda \in \Lambda$; accordingly, we define the function $P : X \to \Xi$, $x \mapsto \xi$, which extracts the first component of x.

The punctuation generates rests as it would in an ordinary English sentence. The two items comprising the function definition are separated by a comma, which would not be necessary in a displayed formula, see (2.16). The formula defining P is echoed by a short sentence, for added clarity.

This is a fully punctuated displayed formula:

$$a_1 = 1; \quad a_{k+1} = \begin{cases} a_k^2 - 1, & 1 \leqslant k < 10; \\ a_k^2, & k \geqslant 10. \end{cases}$$

If full punctuation seems heavy (it may do so here), then we may replace some punctuation by increased spacing or by words:

$$a_1 = 1; \quad a_{k+1} = \begin{cases} a_k^2 - 1 & \text{if } 1 \leqslant k < 10; \\ a_k^2 & \text{if } k \geqslant 10. \end{cases}$$

A displayed formula is normally embedded within a sentence, and the punctuation at the end of a formula must be appropriate to the structure of the sentence. In particular, if a sentence terminates at a formula—as in the example above—then the full stop at the end of the formula must always be present.

6.3 Improving Formulae

EXAMPLE. The following untidy formula

$$f(x) = \frac{14x - 2x^3 - 2x^2 + 14}{-2x - 4}$$

could represent a typical unprocessed output of a computer algebra system. It contains redundant information (a common factor between numerator and denominator), the monomials at numerator are not ordered, and there are too many negative signs. The properties of the rational function $f(x)$ are not evident from it. There are several ways to improve the formula's layout:

$$f(x) = \frac{x^3 + x^2 - 7x - 7}{x + 2}$$

$$f(x) = \frac{(x + 1)(x^2 - 7)}{x + 2}$$

$$f(x) = x^2 - x - 5 + \frac{3}{x + 2}.$$

If the degree or the coefficients of f are important, then the first version is appropriate; the second version makes it easy to solve the equation $f(x) = 0$; the third version is a preparation for integrating f.

EXAMPLE. The following subset of the rationals

$$A = \left\{ y \in \mathbb{Q} : y = \frac{x}{x^2 + 1}, \ x \in \mathbb{Z}, \ x < 0 \right\}$$

is defined with a disproportionate volume of notation. To save symbols we switch from the Zermelo to the standard set definition, and consider only elements of the required form. We also remove the inequality by introducing a negative sign, so we can replace the integers by the natural numbers:

$$A = \left\{ \frac{-n}{n^2 + 1} : n \in \mathbb{N} \right\}.$$

Now the formula is transparent.

EXAMPLE. To simplify a complex formula, we apply the 'divide and conquer' method. In the cluttered expression

$$R = x(ad - bc) - y(ad - cb)^2 + z(ad - cb)^3$$

the sub-expression $ad - bc$ appears as a unit. We exploit this fact to improve the layout:

$$R = x\Delta - y\Delta^2 + z\Delta^3 \qquad \Delta = ad - bc.$$

The formula is tidier, and the structure of R is clearer.

EXAMPLE. Our next challenge is to improve an intricate double sum:

$$z(y_1, y_2, \ldots) = \sum_{i=1}^{\infty} \sum_{y=0}^{y_i-1} (y+1)x^{i-1}. \tag{6.5}$$

The meaning of z is not at all evident. This quantity depends on x, but its dependence does not appear explicitly; a poor choice of symbols obscures matters further. We note that the parameters y_i are integers, since they are the upper limit of the inner summation; accordingly, we replace the symbol y with n—see (6.1). Then we adopt the 'divide and conquer' method, splitting up the sum as follows:

$$z(x, n_1, n_2, \ldots) = \sum_{i=1}^{\infty} d_i x^{i-1} \qquad d_i = \sum_{k=0}^{n_i-1} (k+1).$$

We now see that z is a power series in x; its coefficients are finite sums, determined by the elements of an integer sequence. These are sums of arithmetic progressions, which can be evaluated explicitly:

$$\sum_{0 \leqslant k \leqslant n-1} (k+1) = \sum_{1 \leqslant k \leqslant n} k = \frac{n(n+1)}{2}.$$

(In this passage we have dropped the subscript i, since the association $n \leftrightarrow n_i$ is clear, and we have switched to the standard sigma-notation to change variable—cf. Eq. (3.7)). Our original double sum (6.5) can now be written as

$$z(x, \mathbf{n}) = \frac{1}{2} \sum_{i=1}^{\infty} n_i(n_i+1)x^{i-1} \qquad \mathbf{n} = (n_1, n_2, \ldots)$$

where we have used again the 'divide and conquer' method. The dependence of z on the variable x and the sequence $\mathbf{n}$ is now clear.

6.4 Writing Definitions

A definition requires a pause, to give the reader time to absorb it. This may be achieved by giving the definition twice, first with words then with symbols (or vice-versa), by using two different formulations, or by supporting the definition with an example. Let us analyse some definitions.

EXAMPLE.

Let $\mathscr{P}$ be a predicate over the rational numbers, that is, a function of the type

$$\mathscr{P} : \mathbb{Q} \to \{T, F\}.$$

6.4 Writing Definitions

The definition of the symbol $\mathscr{P}$ uses some jargon (predicate over a set) so the second part of the sentence recalls its meaning. A previous exposure to this concept is tacitly assumed; for a first encounter we would need a more considerate style, such as in the opening paragraph of Sect. 4.3.

EXAMPLE.

For given $\varepsilon > 0$, we define the set

$$S_\varepsilon = \{(x, y) \in \mathbb{R}^2 : |x - y| \leqslant \varepsilon/\sqrt{2}\},$$

that is, S_ε is a strip of width ε symmetrical with respect to the main diagonal in the cartesian plane.

This time the symbolic (Zermelo) definition appears first, while a verbal explanation clarifies the geometric meaning of S_ε, which is not obvious from the formula. The quantity ε appears as a subscript, indicating that it is a parameter. Zermelo definitions should be used sparingly, and should not be used at all if we write for a non-mathematical audience. (Lars Ahlfors in his beautifully written text *Complex analysis* deliberately avoids them [2].)

EXAMPLE.

For every real number λ, let $\Pi(\lambda)$ be the plane in three-dimensional euclidean space orthogonal to the vector $v(\lambda) = (1, \lambda, \lambda^2)$. Thus $\Pi(\lambda)$ consists of the points $z = (x_1, x_2, x_3) \in \mathbb{R}^3$ for which the scalar product $z \cdot v = x_1 + x_2\lambda + x_3\lambda^2$ is equal to zero.

In this definition the quantity λ appears as an argument of both Π and v, and thus the latter represent functions. The second sentence restates the definition in a form suitable to computation, recalling the connection between orthogonality and scalar product, and introducing further notation.

EXAMPLE.

Let $\mathscr{N}$ be the set of sequences of natural numbers, such that every natural number is listed infinitely often. For example, the sequence

$$(1, 1, 2, 1, 2, 3, 1, 2, 3, 4, \ldots)$$

belongs to $\mathscr{N}$.

A non-experienced reader will have no idea that such a construction is at all possible, so we give an example.

The definition of a symbol should appear as near as possible to where the symbol is first used; defining a symbol immediately after its first appearance is also acceptable, provided that the definition is given *within the same sentence.*

EXAMPLE. We give the same definition three times, articulating the changes in emphasis that accompany each version.

Consider the power series

$$h(x) = \sum_{n=1}^{\infty} a_n x^n,$$

where the coefficient a_n is the square of the n-th triangular number.

The definition of a_n immediately follows its first appearance. The displayed formula represents a general power series, and its specific nature is revealed only by reading the entire sentence. For this reason, this format may not be ideal if the formula is to be referred to from elsewhere in the text. This definition puts some burden on readers who are unfamiliar with the term **triangular number**.

In our second version we change both text and formula.

Let t_n be the n-th triangular number. We consider the power series

$$h(x) = \sum_{n=1}^{\infty} t_n^2 x^n.$$

Now t_n is defined before being used, with the symbol t chosen so as to remind us of 'triangular'. The formula is clearer. Just glancing at it makes us want to find out what t_n is, and to do this one would begin to scan the text preceding, rather than following, the formula.

Our third version combines verbal and symbolic definitions.

We consider the power series $h(x)$ whose coefficients are the square of the triangular numbers, namely,

$$h(x) = \sum_{n=1}^{\infty} t_n^2 x^n \qquad t_n = \frac{n(n+1)}{2}.$$

The formula is more complex because everything is defined within it. But it is also self-contained, so it'll be easy to refer to.

6.5 Introducing a Concept

Opening the exposition with a formal definition is rarely a good way to introduce a new concept, particularly if we write for non-experts. The following examples show how to set the scene for a gentle introduction to a new idea. In each case we ask a question—a simple rhetorical device to engage and prepare an audience.

6.5 Introducing a Concept

EXAMPLE. Introducing recursive sequences.

Let n be a natural number. How is the power 2^n defined? We could use repeated multiplication

$$2^n := \underbrace{2 \times 2 \times \cdots \times 2}_{n},$$

but we could also write

$$2^1 := 2 \quad \text{and} \quad 2^n := 2 \times 2^{n-1} \quad n > 1.$$

The second formula is an example of a **recursive definition** *of a sequence (a_n). When $n = 1$, the first term $a_1 = 2$ is defined explicitly; then, assuming that a_{n-1} has been defined, we define a_n in terms of it.*

[The recursive definition of a general sequence follows (see Sect. 9.4)].

The opening question leads to the familiar definition of exponentiation in terms of multiplication. The less familiar recursive definition comes after, illustrated by a simple example and supported by an appropriate notation. After this preamble, the reader should be ready to confront an abstract definition.

EXAMPLE. Introducing the exponential function.

The process of differentiation turns a real function into another real function. For example, differentiation turns the sine into the cosine. Are there functions that are not changed at all by differentiation?

[The definition of the exponential function follows.]

We recall a structural property of differentiation and present a familiar example. The question that follows suggests how the argument will develop.

EXAMPLE. Introducing the rational numbers.

A rational number is represented by a pair of integers, the numerator and the denominator. As these integers need not be co-prime, we may choose them in infinitely many ways. How are we to construct a single rational number from an infinite set of pairs of integers? How do we define the set $\mathbb{Q}$ from the cartesian product $\mathbb{Z} \times (\mathbb{Z} \setminus \{0\})$?

[The definition of the set of rational numbers follows.]

The first two sentences recall elementary facts. A question then invites the reader to think about this problem more carefully. A second question, which echoes the first, uses proper terminology and notation, in preparation for a formal construction.

EXAMPLE. We design an exercise structured as a list of questions. The topic is the number of relations on a finite set.

What is a relation on a set?
Can a relation be defined on the empty set?
How many relations can one define on a two-element set?
Let n be the number of relations on a set. What values can n assume?

The first question checks background knowledge; the other questions gently explore the problem, guiding the reader from the specific to the general. Textbook exercises are sometimes structured in this way, to encourage independence in learning.

Formulating questions is not just a method to grab attention or structure exercises. Asking the right questions—those that chart the boundaries of our knowledge—is the essence of research.

6.6 Writing a Short Description

Writing a synopsis of a mathematical topic is a common task. It could be the abstract of a presentation, the summary of a chapter of a book, an informal explanation of a theorem. It could be the closing paragraph of a large document, which distils the essence of an entire subject.

Writing a short essay is difficult—the shorter the essay, the greater the difficulty. We shall adopt the format of a MICRO-ESSAY: 100–150 words (one or two paragraphs) and no mathematical symbols. Within such a confined space one is forced to make difficult decisions on what to say and what to leave out; the lack of symbols gives further prominence to the concepts. Our command of the syntax will be put to the test.

Our first MICRO-ESSAY is a summary of Sect. 2.1 on sets. We have access to all relevant material, and the main difficulty is to decide what are the highlights of that section. We select two ideas: how to define a set, and how to construct new sets from old ones.

> *A set is a collection—finite or infinite—of distinct mathematical objects, whose ordering is unimportant. A small set may be defined by listing all its elements explicitly; a large set is instead defined by specifying the characteristic properties of the elements.*
>
> *Combining numbers with arithmetical operators gives new numbers. Likewise, combining sets via set operators (union, intersection, difference) gives new sets. But this is a bit like recycling what we already have. A class of brand-new sets are the so-called cartesian products, which are constructed by pairing together elements of existing sets. A well-known example is the cartesian plane, which is made of pairs of elements of the real line, each representing a coordinate.*

The description of set operators exploits an analogy with arithmetical operators, and for this reason the first two sentences of the second paragraph have the same structure. We avoid a precise definition of the term 'cartesian product' (not enough space!), opting instead for an informal description and an illustrative example.

6.6 Writing a Short Description

Next we write a MICRO-ESSAY on prime numbers, a synthesis of our knowledge of this topic.

A prime is a positive integer divisible only by itself and unity. (However, 1 is not considered prime). The importance of primes in arithmetic stems from the fact that every integer admits a unique decomposition into primes. The infinitude of primes—known from antiquity—and their unpredictability make them object of great mathematical interest.

Many properties of an integer follow at once from its prime factorization. For instance, looking at the exponents alone, one can determine the number of divisors, or decide if an integer is a power (i.e.,[6] a square or a cube).

Primality testing and prime decomposition are computationally challenging problems with applications in digital data processing.

This essay begins with a definition. The technical point concerning the primality of 1 has been confined within parentheses, to avoid cluttering the opening sentence. The following two sentences deliver core information in a casual—yet precise—way. We state two important theorems (the Fundamental Theorem of Arithmetic, and Euclid's theorem on the infinitude of the primes) without mentioning the word 'theorem'. We also give a hint of why mathematicians are so fascinated by primes. The second paragraph elaborates on the importance of unique prime factorisation, by mentioning two applications without unnecessary details. The short closing paragraph, like the last sentence of the first paragraph, is an advertisement of the subject, meant to encourage the reader to learn more.

Now a real challenge: write a MICRO-ESSAY on Theorem 4.2, which we reproduce here for convenience.

Theorem. *Let X be a set, let $A, B \subseteq X$, and let $\mathscr{P}_A$, $\mathscr{P}_B$ be the corresponding characteristic functions. The following holds (the prime denotes taking complement):*

(i) $\quad\quad\quad\quad\neg\ \mathscr{P}_A\ =\ \mathscr{P}_{A'}$
(ii) $\quad\ \mathscr{P}_A\ \wedge\ \mathscr{P}_B\ =\ \mathscr{P}_{A \cap B}$
(iii) $\quad\ \mathscr{P}_A\ \vee\ \mathscr{P}_B\ =\ \mathscr{P}_{A \cup B}$
(iv) $\quad\ \mathscr{P}_A\ \Rightarrow\ \mathscr{P}_B\ =\ \mathscr{P}_{(A \smallsetminus B)'}$
(v) $\quad\ \mathscr{P}_A\ \Longleftrightarrow\ \mathscr{P}_B\ =\ \mathscr{P}_{(A \cap B) \cup (A \cup B)'}$.

We must extract a theme from a daunting list of inscrutable symbols. We inspect the material of Sect. 4.3 leading to the statement of this theorem: it deals with characteristic functions, and the main idea is to link characteristic functions to sets. We can see such a link in every formula: on the left there are logical operators, on the right set operators. Given the specialised nature of this theorem, some jargon is unavoidable, so we write for a mathematically mature audience.

Every characteristic function corresponds to a set, and vice-versa. If we let a logical operator act on characteristic functions we obtain a new characteristic

---

[6] Abbreviation for the Latin id est, meaning 'that is'.

function, and with it a new set. How is this set related to the original sets? This theorem tells us that for each logical operator there is a set operator which mirrors its action on the corresponding sets. In other words, we have a bi-unique correspondence between these two classes of operators. For instance, the logical operator of negation is represented by the set operation of taking the complement. More precisely, the negation of the characteristic function of a set is the characteristic function of the complement of this set. Analogous results are established with respect to the other logical operators.

Here words are better than symbols to describe the structure, but they can't compete with symbols for the details. So, to avoid tedious repetitions, we have chosen to explain just one formula carefully (the easiest one!), mentioning the other formulae under the generic heading 'analogous results'.

Exercise 6.1 Answer concisely. [$\not{\in}$, 30]

1. What is the difference between an ordered pair and a set?
2. What is the difference between an equation and an identity?
3. What is the difference between a function and its graph?

Exercise 6.2 Consider the following question:

Why is it that when the price of petrol goes up by 10 % and then comes down 10 %, it doesn't finish up where it started?

1. Write an explanation for the general public. Do not use mathematical symbols, as most people find them difficult to understand. [$\not{\in}$]
2. Write an explanation for mathematicians, combining words and symbols for maximum clarity. You should deal with the more general problem of two opposite percentage variations of an arbitrary non-negative quantity, not specialised to petrol, or to 10 % variation, or to the fact that the decrease followed the increase and not the other way around.

Exercise 6.3 Consider the following question:

I drive ten miles at 30 miles an hour, and then another ten miles at 50 miles an hour. It seems to me my average speed over the journey should be 40 miles an hour, but it doesn't work out that way. Why not?

Write an explanation for the general public, clarifying why such a confusion may arise. You may perform some basic arithmetic, but do not use symbols as most people find them difficult to understand. [$\not{\in}$]

Exercise 6.4 Consider the following question:

I tossed a coin four times, and got heads four times. It seems to me that if I toss it again I am much more likely to get tails than heads, but it doesn't work out that way. Why not?

6.6 Writing a Short Description

Write an explanation for the general public, clarifying why such a confusion may arise. You may use symbols such as H and T for heads-tails outcomes, but avoid using other symbols. [✗]

Exercise 6.5 Consider the following question:

> In a game of chance there are three boxes: two are empty, one contains money. I am asked to choose a box by placing my hand over it; if the money is in that box, I win it. Once I have made my choice, the presenter—who knows where the money is—opens an empty box and then gives me the option to reconsider. I can change box if I wish. It seems to me that changing box would make no difference to my chances of winning, but it does not work out that way. Why not?

Write for the general public, explaining what is the best winning strategy. [✗]

Exercise 6.6 Write a MICRO-ESSAY on each topic. [✗, 150]
1. Quadratic equations and complex numbers.
2. Definite versus indefinite integration.
3. What are matrices useful for?
4. Images and inverse images of sets.
5. Rational numbers versus fractions.
6. How to get the integers from pairs of natural numbers.

Exercise 6.7 Write a two-page essay on each topic.
1. The many ways of defining the logarithm.
2. All fractions in a very small interval have very large denominators (with at most one exception, that is).
3. In probability, a random variable is not random and is not a variable: it's a function!
4. Sequences racing to infinity: who gets there first?

Chapter 7
Forms of Argument

A mathematical theory begins with a collection of **axioms** or **postulates**. These are statements that are assumed to be true—no verification or justification is required. The axioms are the building blocks of a theory: from them we deduce other true statements called **theorems**, and a **proof** is the process of deduction that establishes a theorem. As more and more theorems are proved, the theory is enriched by a growing list of true statements.

When we develop a mathematical argument, we need to organise sentences so that every sentence is either an axiom or a true statement derived from axioms or earlier statements. In practice only significant statements deserve to be called theorems. In the proof of a theorem many true statements may be derived, but these fragments are not usually assigned any formal label. Sometimes a proof rests on statements which, if substantial, are not of independent significance: they are called **lemmas**. Finally, a **proposition** is a statement which deserves attention, but which is not sufficiently general or significant to be called a theorem.

In this chapter we survey some methods to give shape to a mathematical argument. Our survey will continue in Chap. 8, where we consider induction techniques.

7.1 Anatomy of a Proof

A first analysis course begins with a list of axioms defining the real number system. Then one may be asked to prove that

$$\forall x \in \mathbb{R}, \quad -(-x) = x. \tag{7.1}$$

But isn't this statement obvious? In recognising the truth of this assertion we make implicit use of knowledge derived from experience, which is not part of the axioms. A proof from the axioms requires erasing all previous knowledge.

We now introduce some axioms, state a theorem equivalent to (7.1), and then prove it from the axioms. We put the proof under X-ray. Our purpose is to dissect a mathematical argument articulating every step; we use the language mechanically, to facilitate the identification of the logical elements of the proof. To avoid making implicit assumptions, we represent familiar objects with an unfamiliar notation.

We are given a set Ω and a binary operator '$\odot$' on Ω, with the following properties:

G1: $\forall x, y, \in \Omega, \ x \odot y \in \Omega$
G2: $\forall x, y, z \in \Omega, \ (x \odot y) \odot z = x \odot (y \odot z)$
G3: $\exists \diamond \in \Omega, \ \forall x \in \Omega, \ x \odot \diamond = \diamond \odot x = x$
G4: $\forall x \in \Omega, \ \exists x' \in \Omega, \ x \odot x' = \diamond$

(These axioms define a general object: a **group**.) Setting $\Omega = \mathbb{R}$, and '$\odot$' = '+', one recognises that $\diamond$ represents 0 and x' represents $-x$. We are ready to state and prove our theorem.

1. **Theorem**.
2. $\forall x \in \Omega, \ x'' = x$.
3. PROOF.
4. Let $x \in \Omega$ be given;
5. then (by G4) $x'' \in \Omega$;
6. let $a := (x \odot x') \odot x''$, $b := x \odot (x' \odot x'')$;
7. hence (by G1 and 5) $a, b, \in \Omega$, and (by G2) $a = b$;
8. then (by G3 and G4) $a = \diamond \odot x'' = x''$;
9. then (by G3 and G4) $b = x \odot \diamond = x$;
10. hence (by 7, 8, 9) $x'' = x$.
11. $\square$

Items 1, 3, 11 are **logical tags**—expressions or symbols that say things about the text. They announce the statement of the theorem, and the beginning and the end of the proof, respectively. The symbol '$\square$' in item 11 may be replaced by the acronym Q.E.D.[1]

Items 5, 8, 9 begin with the logical tag 'then', used here to mean that what comes after is deduced from the axioms. Formally, item 5 is our first theorem, being a true statement deduced from the axioms. Items 7 and 10 are deductions of a slightly different nature, and we have flagged them with the tag 'hence'. These items use not only axioms but also facts assembled from previous statements.

Items 4 and 6 are not statements but **instructions**. Item 4 instructs us to assume that x is an arbitrary element of a set. This is a standard opening sentence which mirrors the expression $\forall x \in \Omega$ in the theorem. Item 6 instructs us to define two new quantities. Many different instructions may be found in proofs, such as the instruction to draw a picture.

---

[1] Abbreviation for the Latin *quod erat demonstrandum*, meaning 'which had to be demonstrated'.

7.1 Anatomy of a Proof

In item 10 we use implicitly the **transitivity** of the equality operator (if $x = y$ and $y = z$, then $x = z$), which depends on the fact that equality is an equivalence relation (see Sect. 4.6).

Proofs from axioms are necessarily rare. Our next proof—the irrationality of $\sqrt{2}$—is quite removed from the axioms of arithmetic, as it relies on several definitions and facts without justification. In this respect this proof is more typical than the previous one, even though the style and conventions are the same.

1. **Theorem.**
2. $\sqrt{2}$ is irrational.
3. PROOF.
4. Suppose $\sqrt{2}$ is rational;
5. then $\exists m, n \in \mathbb{Z}$ such that $\sqrt{2} = m/n$.
6. Choose m, n in 5 so that they are co-prime.
7. Since $\sqrt{2} = m/n$,
8. then $2 = m^2/n^2$;
9. then $2n^2 = m^2$;
10. then m^2 is even;
11. hence (by 10 and Theorem A[2]) m is even;
12. then $m = 2h$ for some integer h;
13. hence (by 9 and 12) $2n^2 = 4h^2$;
14. then $n^2 = 2h^2$;
15. then n^2 is even;
16. hence (by 15 and theorem A) n is even;
17. hence (by 11 and 16) m and n are not co-prime;
18. hence (by 6 and 17) assumption 4 is false;
19. then its negation is true.
20. □

A fair amount of mathematics is assumed in the proof:

> the definition of the rational numbers and of $\sqrt{2}$ (items 2, 4, 5);
> the definition of co-primality (item 6);
> some properties of equations (items 8, 9);
> the definition of an even integer (items 12, 17);
> a theorem of arithmetic, referred to as Theorem A (items 11, 16).

The argument contains novel elements. The core of the proof is item 4, the **assumption** that $\sqrt{2}$ is rational. This is puzzling: how can we assume that $\sqrt{2}$ is rational when we know it isn't? How can we let $\sqrt{2} = m/n$ when we haven't been told what numbers m and n are? This is not so strange; assumptions belong very much to common reasoning.

Suppose that I meet Einstein on top of Mount Everest.
Suppose that people walk laterally.

---

[2] If a prime divides the product of two integers, then it divides one of the factors.

We are clearly free to explore the logical consequences of these assumptions. In mathematics, assumptions are handled formally via the implication operator ⇒ developed in Chap. 4.

The assumption that $\sqrt{2}$ is rational eventually leads to a **contradiction**: items 6 and 17 are conflicting statements. From this fact we deduce that the assumption in item 4 is false (item 18). Here we abandon the assumption 4. But if the statement $\sqrt{2} \in \mathbb{Q}$ is false, then its negation is true. This is item 19, which is what we wanted to prove.

We now abandon the robotic, over-detailed proof style adopted in this section, and analyse various methods of proof in more typical settings.

7.2 Proof by Cases

Some mathematical arguments are made tidier by breaking them into a number of cases, of which precisely one holds, while all lead to the desired conclusion.

EXAMPLE. Consider the following statement:

The solution set of the inequality $2|x| \geqslant |x-1|$ is the complement of the open interval $(-1, 1/3)$.

PROOF. Let x be a real number. There are three cases.

(1) If $x < 0$, then the inequality is $-2x \geqslant 1-x$, giving $x \leqslant -1$.
(2) If $0 \leqslant x < 1$, then the inequality is $2x \geqslant 1-x$, giving $1/3 \leqslant x < 1$.
(3) If $x \geqslant 1$, then the inequality is $2x \geqslant x-1$, giving $x \geqslant 1$.

So the required solution set is the collection of the values of x such that $x \leqslant -1$, or $1/3 \leqslant x < 1$, or $x \geqslant 1$, which is the union of two rays:

$$(-\infty, -1] \cup [1/3, 1) \cup [1, \infty) = (-\infty, -1] \cup [1/3, \infty).$$

This set is obtained from the real line by removing the open interval $(-1, 1/3)$, as claimed. □

The opening sentence 'Let x be a real number' acknowledges that all real values of x will have to be considered. The second sentence announces that the proof will branch into three cases, determined by the sign of the expressions within absolute value. Note the careful distinction between strict and non-strict inequalities, to avoid missed or repeated values of x.

The structure of this proof is shaped by the structure of the absolute value function, which is piecewise-defined—see Sect. 5.6. The presence of piecewise-defined functions usually leads to proofs by cases.

EXAMPLE. A proof by cases on divisibility.

Let n be an integer; then $n^5 - n$ is divisible by 30.

7.2 Proof by Cases

PROOF. We factor the integer 30 and the polynomial $n^5 - n$:

$$30 = 2 \times 3 \times 5 \qquad n^5 - n = n(n-1)(n+1)(n^2+1).$$

For each prime $p = 2, 3, 5$ we will show that, for any integer n, at least one factor of $n^5 - n$ is divisible by p.

1. $p = 2$. Let $n = 2k + j$, for some $k \in \mathbb{Z}$ and $j = 0, 1$.
   If $j = 0$, then n is divisible by 2.
   If $j = 1$, then $n - 1$ is divisible by 2.
2. $p = 3$. Let $n = 3k + j$, for some $k \in \mathbb{Z}$ and $j = 0, \pm 1$.
   If $j = 0$, then n is divisible by 3;
   if $j = \pm 1$, then $n \mp 1$ is divisible by 3.
3. $p = 5$. Let $n = 5k + j$, for some $k \in \mathbb{Z}$ and $j = 0, \pm 1, \pm 2$.
   If $j = 0$, then n is divisible by 5.
   If $j = \pm 1$, then $n \mp 1$ is divisible by 5.
   If $j = \pm 2$, then $n^2 + 1 = 25k^2 \pm 20k + 5 = 5(5k^2 \pm 4k + 1)$ is divisible by 5.

The proof is complete. □

After the factorisations we declare our intentions. There are three cases, one for each prime divisor p of 30. Each prime in turn leads to p cases, corresponding to the possible values of the remainder j of division by p. To simplify book-keeping, we consider also negative remainders, and then pair together the remainders which differ by a sign. Note the presence of the symbols $\pm$ and $\mp$ within the same sentence. The positive sign in the first expression matches the negative sign in the second expression, and vice-versa.

7.3 Implications

Many mathematical statements take the form of an implication:

> If P then Q.

For example:

> If p is an odd prime, then $2^{p-1} - 1$ is divisible by p.
> If the sequence (a_k) is periodic, then (a_{2k}) is also periodic.

Implications may appear in disguise, without the 'if…then' construct:

> A is a subset of B.
> Every repeating decimal is rational.
> The determinant of an invertible matrix is non-zero.

When we rewrite these sentences as explicit implications, we note that it is necessary to introduce an auxiliary quantity:

If $x \in A$, *then* $x \in B$.
If r is a repeating decimal, then r is rational.
If A is an invertible matrix, then $\det(A) \neq 0$.

This is due to the presence of a hidden universal quantifier. So the symbolic form of an implication is not $P \Rightarrow Q$, but rather

$$\forall x \in X, \quad \mathscr{P}(x) \Rightarrow \mathscr{Q}(x) \tag{7.2}$$

where $\mathscr{P}$ and $\mathscr{Q}$ are predicates over X. Let us expand one of our sentences so as to tell the full story:

For all real numbers r, if the decimal digits of r are repeating, then r is rational.

Spelling out an implication in this way will help us structure its proof.

7.3.1 Direct Proof

From the truth table (4.9) we see that if P is false, then $P \Rightarrow Q$ is true regardless of the value of Q, so there is nothing to prove. If P is true, then the implication is true provided that Q is true. So a **direct proof** of the implication consists in **assuming** P and **deducing** Q. The assumption of P lasts only until we have proved Q; after that P is discharged and may no longer be used.

Every direct proof of '*If P then Q*' must contain a section during which P is assumed. The beginning of the section is announced by a sentence such as

   Assume P.   *Suppose P.*   *Let P.*

The task of the rest of the proof is to prove not the original theorem, but Q. We make this clear by writing

   RTP: Q

where RTP stands for *Remains To Prove* (or *Required To Prove*). The block ends when the proof of Q is completed. This is announced with a closing sentence such as

   We have proved Q.   *The proof of Q is complete.*

The above considerations suggest what should be the opening sentence of a direct proof of an implication:

Theorem. *If $p > 3$ is a prime and $p+2$ is also prime, then $p+4$ is composite.*

PROOF. Suppose p is a prime number greater than 3, such that $p + 2$ is prime.
RTP: $p + 4$ is composite.

7.3 Implications

All three assumptions ($p > 3$, p prime, $p + 2$ prime) must be used in the proof. If not, either the proof is wrong or some assumption is redundant.

Often you can work out how the proof of an implication must start even if you haven't the faintest idea of what the mathematics is about. We illustrate this point with some real life examples:

> **Theorem.** *A closed subset of a compact set is compact.*
> PROOF. Let X be a compact set, and let C be a subset of X. Assume that C is closed. RTP: C is compact.
> **Theorem.** *If $\lambda \in \mathbb{C}$ is a root of a monic polynomial whose coefficients are algebraic integers, then λ is an algebraic integer.*
> PROOF. Let p be a monic polynomial whose coefficients are algebraic integers, and let $\lambda \in \mathbb{C}$ be a root of p. RTP: λ is an algebraic integer.
> **Theorem.** *Every basis of a finite-dimensional linear space has the same number of elements.*
> PROOF. Let V be a finite-dimensional linear space, and let B_1 and B_2 be two bases for V. Suppose B_1 has n_1 elements and B_2 has n_2 elements. RTP: $n_1 = n_2$.

Establishing a good notation is often decisive. In all the examples above, the proof begins by giving names to things. Some authors make this unnecessary by including names in the theorem; others obscure the statement of a theorem by putting too many names in it. Let us rewrite the last theorem in such a way as to establish some notation within the statement itself.

> **Theorem.** *Let V be a finite-dimensional linear space. Then every basis for V has the same number of elements.*
> **Theorem 1.** *Let V be a finite-dimensional linear space, and let B_1 and B_2 be two bases for V. Then $\#B_1 = \#B_2$.*

Let us compare the three formulations of this theorem. The first is plain and effective. The second contains a minimum of notation, which does no harm but is unnecessary. The last version establishes some useful notation. Normally the notation should be introduced within the proof, unless one plans to use it at a later stage. For instance, one could find the sentence

> Let V, B_1, B_2 be as in Theorem 1.

in the text following the theorem.

7.3.2 Proof by Contrapositive

In Sect. 4.2 we saw that every implication $P \Rightarrow Q$ is equivalent to its **contrapositive** $\neg Q \Rightarrow \neg P$: they are both true or both false for any value of P and Q. This equivalence gives us a method for proving implications, called **proof by contrapositive**, which is a useful alternative to a direct proof. To prove by contrapositive that

If P then Q,

we prove instead that

If not Q then not P,

that is, we assume that Q is false and then deduce that P is false.

A proof by contrapositive is structurally identical to a direct proof, only the predicates are different. Thus in place of (7.2) we consider the statement

$$\forall x \in X, \quad \neg \mathcal{Q}(x) \Rightarrow \neg \mathcal{P}(x). \tag{7.3}$$

In forming the contrapositive we haven't altered the quantifier; we have merely replaced a boolean function with an equivalent function, much like replacing $\sin^2(x)$ with $1 - \cos^2(x)$.

How do we decide between a direct proof and a proof by contrapositive? We must compare the assumptions P and $\neg Q$, and decide which of the two is easier to handle. Sometimes it's necessary to try both approaches to find out. The following examples show the reasoning behind such decisions.

EXAMPLE. Consider the statement

$$\forall n \in \mathbb{N}, \quad (2^n < n!) \Rightarrow (n > 3). \tag{7.4}$$

The assumption $\mathcal{P}(n) = (2^n < n!)$ is problematic, because its value is not easily computable. By contrast, $\neg \mathcal{Q}(n) = (n \leqslant 3)$ is straightforward, and the contrapositive implication

$$\forall n \in \mathbb{N}, \quad n \leqslant 3 \Rightarrow (2^n \geqslant n!) \tag{7.5}$$

involves checking that the boolean expression $(2^n \geqslant n!)$ is true for only three values of n.

PROOF. We only have to check three cases:

$n = 1$: $2 = 2^1 \geqslant 1! = 1$
$n = 2$: $4 = 2^2 \geqslant 2! = 2$
$n = 3$: $8 = 2^3 \geqslant 3! = 6$.

Thus the expression (7.5) is true, and the proof is complete. □

EXAMPLE. Consider the statement

If the average of four distinct integers is equal to 10, then one of the integers is greater than 11.

The direct implication involves an assumption on an average value, which entails loss of information; the contrapositive implication involves four integers of bounded size. We opt for the latter, which seems easier.

Given four distinct integers not greater than 11, their average is not equal to 10.

PROOF. Let four distinct integers be given. If none of them exceeds 11, then the largest value their sum can assume is $11 + 10 + 9 + 8 = 38$. So the largest possible average is
$$\frac{38}{4} = \frac{19}{2} < 10$$
as desired. □

7.4 Conjunctions

A **conjunction** is a statement of the type

P and Q.

The statements P and Q are called the **conjuncts**. As with implications, conjunctions are not necessary presented explicitly.

A direct proof of a conjunction consists of the separate proofs of the two conjuncts, which must be differentiated clearly. It's common practice to put the conjuncts in an ordered list, say (i) and (ii). (If there are more than two conjuncts, the list may be extended using lower-case Roman numerals: (iii), (iv), etc.) Then the proof itself should use the matching labels (i) and (ii), and in each case we should begin by stating what we intend to prove.

PROOF.

(i) We prove P. ...
(ii) We prove Q. ...

Equality of sets is the canonical hidden conjunction:

$A = B.$

What are the conjuncts? Two sets are equal if they have the same elements, namely every element of A is an element of B, and vice-versa.

$$(A = B) \Leftrightarrow (A \subset B) \wedge (B \subset A).$$

The structure of the proof is now determined by the definition (4.19) of a subset.

PROOF.

(i) We prove that $A \subset B$. Let $x \in A$ be given. RTP: $x \in B$.
(ii) We prove that $B \subset A$. Let $x \in B$ be given. RTP: $x \in A$.

Another basic conjunction is the equivalence of two statements:

P if and only if Q.

The equivalence operator $\Leftrightarrow$ is the conjunct of an implication and its converse—see (4.12).

PROOF.

(i) We prove $P \Rightarrow Q$. ...
(ii) We prove $Q \Rightarrow P$. ...

We could replace either part by a proof of the contrapositive implication.

The following theorem is of this type: it provides an alternative characterisation of primality.

Theorem. [3] *A natural number $n > 1$ is prime if and only if n divides $(n-1)! + 1$.*

This theorem says $P \wedge Q$, where

$$P = \forall n > 1, \ (n \text{ is prime}) \Rightarrow (n \mid (n-1)! + 1)$$
$$Q = \forall n > 1, \ (n \mid (n-1)! + 1) \Rightarrow (n \text{ is prime}).$$

The outline of the proof is now clear.

PROOF.

(i) Let n be a prime number.
   RTP: n divides $(n-1)! + 1$.
(ii) Let n be a natural number greater than 1 which divides $(n-1)! + 1$.
   RTP: n is prime.

The adverb 'precisely' may be used to turn a one-sided implication into an equivalence, hence a conjunction. Thus the proof of the statement

The set $\mathbb{Z}/n\mathbb{Z}$ is a field precisely if n is prime.

will have to be carried out in two stages.

PROOF. Let $n \in \mathbb{N}$ be given.

(i) Assume that $\mathbb{Z}/n\mathbb{Z}$ is a field. RTP: n is prime.
(ii) Assume that n is prime. RTP: $\mathbb{Z}/n\mathbb{Z}$ is a field.

7.4.1 Loops of Implications

The equivalence of several statements is sometimes expressed by a chain of implications in which the last statement in the chain coincides with the first one, thus forming a loop. This type of argument combines conjunctions with implications.

The equivalence of two statements P_1 and P_2, may be viewed as a loop:

$$P_1 \Rightarrow P_2 \Rightarrow P_1.$$

---

[3] This result, known as Wilson's theorem, was first formulated by Bhaskara (a 7th Century Indian mathematician) and first proved by Lagrange.

7.4 Conjunctions

More generally, we consider loops of n implications:

$$P_1 \Rightarrow P_2 \Rightarrow \cdots \Rightarrow P_{n-1} \Rightarrow P_n \Rightarrow P_1.$$

Proving all implications in the loops amounts to proving that all statements are equivalent, namely that $P_i \Leftrightarrow P_j$, for all $i, j = 1 \ldots, n$. This follows from the transitivity of the implication operator.

For example, let G be a group and H a subgroup of G. A **left coset** of H in G is a set $gH = \{gh : h \in H\}$ where g is an element of G. [This is a variant of the algebraic product of sets (2.21)]. A result in group theory states that if x, y are elements of G, then the following statements are equivalent:

(a) $xH = yH$
(b) $xH \subseteq yH$
(c) $xH \cap yH \neq \emptyset$
(d) $y^{-1}x \in H$.

These are usually proved in a circle, for example $(a) \Rightarrow (b) \Rightarrow (c) \Rightarrow (d) \Rightarrow (a)$, but other arrangements are possible, e.g., $(a) \Leftrightarrow (b) \Rightarrow (c) \Rightarrow (d) \Rightarrow (b)$.

In Chap. 8 we will use a loop of implications to show the equivalence of four formulations of the principle of induction.

7.5 Proof by Contradiction

A proof by contradiction of a proposition P consists of assuming $\neg P$ and deducing a false statement:

$$\neg P \Rightarrow \text{FALSE}. \tag{7.6}$$

The false statement could be anything, including any assertion that contradicts the assumption $\neg P$. Formally, contradiction works because from (7.6) and the truth table (4.9) of the implication operator we deduce that $\neg P$ is false and hence P is true.

For example, in the proof by contradiction of the irrationality of $\sqrt{2}$, presented in Sect. 7.1, we assumed that $\sqrt{2} \in \mathbb{Q}$, and deduced that

$$\exists m, n \in \mathbb{N}, \ (\gcd(m, n) = 1) \Rightarrow (m, n \in 2\mathbb{Z})$$

which is clearly false.

A classic proof by contradiction is Euclid's proof of the infinitude of primes.

Theorem 7.1. *The number of primes is infinite.*

PROOF. Assume there are only finitely many primes, $p_1, \ldots, p_n$. Consider the integer

$$N = 1 + \prod_{k=1}^{n} p_k. \tag{7.7}$$

Then N is greater than all the primes; moreover, if we divide N by p_k we get remainder 1, and therefore N is not divisible by any of the primes. Now, every integer greater than 1 is divisible by some prime. It follows that N must be divisible by a prime not in the given list, a contradiction. □

The statement that every integer greater than 1 is divisible by some prime needs some justification (see Sect. 8.1).

To apply the method of contradiction to an implication $P \Rightarrow Q$, we assume the negation of the implication, namely $\neg(P \Rightarrow Q)$. By Theorem 4.1, this is equivalent to $P \wedge \neg Q$, so our proof amounts to establishing that

$$(P \wedge \neg Q) \Rightarrow \text{FALSE}.$$

This form of proof by contradiction is called the **both ends method**: to prove that *if P then Q*, we assume both P and not-Q, and we deduce something impossible. The appealing feature of this method is that it gives us two assumptions to exploit, namely P and $\neg Q$. This is often how mathematicians work, although this strategy may not be made explicit. It also easily causes confusion in writing, as the following example (from a textbook) shows.

Theorem. *For all real numbers x, if $x > 0$, then $1/x > 0$.*

BAD PROOF: If $1/x < 0$ then $(-1)/x > 0$, so $x((-1)/x) > 0$, i.e., $-1 > 0$, a contradiction. Therefore $1/x \geqslant 0$. Since $1/x$ can't be 0, we conclude that $1/x > 0$. □

In this proof, the assumption of P is hidden, and the assumption of $\neg Q$ is confused. Not helpful writing!

PROOF. We use contradiction. Let us assume that $x > 0$ and $1/x \leqslant 0$. Multiplying the second inequality by -1, we obtain $-1/x \geqslant 0$, and since x is positive, multiplication by x yields $x(-1/x) \geqslant 0$. Simplification gives $-1 \geqslant 0$, which is false. □

The final deduction rests on the inequality $0 < 1$, which may be derived from the axioms of the real number system.

7.6 Counterexamples and Conjectures

Suppose that a statement concerning all elements of a set is false. To prove this, it is sufficient to exhibit a single element of that set for which the statement fails. This argument is called a **counterexample**. In symbols, we show that the statement

$$\forall x \in X, \ \mathscr{P}(x)$$

is false by proving its negation, namely

$$\exists x \in X, \ \neg \mathscr{P}(x).$$

7.6 Counterexamples and Conjectures

To explore the idea of a counterexample, let us consider a famous polynomial proposed by Euler: $p(n) = n^2 + n + 41$. We evaluate $p(n)$ at integer values of the indeterminate: $n = 0, 1, \ldots$.

n	0	1	2	3	4	5	6	7	8	9	10	$\cdots$	20	$\cdots$	30
$p(n)$	41	43	47	53	61	71	83	97	113	131	151	$\cdots$	461	$\cdots$	971

All these values of $p(n)$ are prime! Moreover, $p(-n) = p(n-1)$, so it seems plausible—if a bit daring—to put forward the following conjecture:

Conjecture 1. *For all integers n, $p(n)$ is prime.*

A conjecture is a statement that we wish were a theorem. Three things may happen to a conjecture: (i) someone proves it, and the conjecture becomes a theorem; (ii) someone produces a counterexample, and the conjecture is proved false; (iii) none of the above, and the conjecture remains a conjecture.

Our case is (ii). The negation of conjecture 1 is

There is an integer n such that $p(n)$ is composite.

To prove it, we exhibit such a value of n. For $n = 40$, we find

$$p(40) = 40^2 + 40 + 41 = 40(40 + 1) + 41 = 41(40 + 1) = 41^2.$$

So $p(40)$ is not prime, and conjecture 1 is false.

For a mathematician, the formulation of a conjecture is accompanied by the fear of a counterexample. Let us return to conjecture 1: we know it's false, but can it be modified into a weaker statement? It can be verified that $n = 40$ is the smallest value for n for which $p(n)$ is not prime.[4] Are there other such values? If these values were exceptional, then a meaningful weaker version of conjecture 1 could be

Conjecture 2. *For all but finitely many integers n, $p(n)$ is prime.*

Even if $p(n)$ were composite for, say, a billion values of n, the conjecture would still hold. Unfortunately, conjecture 2 is not true either. Its negation

There are infinitely many integers n for which $p(n)$ is composite.

is established by the simple, devastating counterexample:

$$p(41k) = (41k)^2 + 41k + 41 = 41(41k^2 + k + 1) \qquad k = 1, 2, \ldots \qquad (7.8)$$

which says that 41 is a proper divisor of $p(41k)$ for infinitely many values of k.

Let's make one final attempt to salvage something from our original claim.

Conjecture 3. *There are infinitely many integers n such that $p(n)$ is prime.*

---

[4] This phenomenon has deep roots, see [9, p. 155].

This is a very meaningful conjecture. Nobody has ever been able to prove that any quadratic polynomial with integer coefficients assumes infinitely many prime values. On the other hand, if we perform a numerical experiment we discover that, as n increases, $p(n)$ continues to supply prime values, even though the composite values slowly become dominant (see Exercise 10.4.1). So this conjecture is very likely to be true, and it makes sense to put it forward.

Arguably, the most famous conjecture in mathematics is the **Riemann's hypothesis** (RH), formulated in 1859. This conjecture (essentially) says that the zeta-function defined in Eq. (6.3), vanishes only at points whose real part is equal to $1/2$. The depth of this statement is not at all apparent from its idiosyncratic formulation.

In a sense, the mathematics community cannot afford to wait for this important matter to be settled, and several theorems about the RH have been proved. Some of these theorems formulate the RH in terms of equivalent statements, establishing the importance of the RH in many areas of mathematics. There are also proofs of **conditional theorems**, which assume the validity of the RH, and deduce some of its consequences.

Other famous conjectures include the **twin-primes conjecture**, attributed to Euclid:

There are infinitely many primes p such that $p + 2$ is also prime.

and the **Goldbach conjecture**[5]:

Every even integer greater than 2 can be written as a sum of two primes.

Both conjectures are easy to formulate and widely believed to be true, but their proof seems hopelessly difficult. This situation is common in the theory of numbers, where assessing the difficulty of a problem is not always easy. To illustrate this phenomenon, let us arrange the natural numbers into four columns, with the primes highlighted in boldface:

$$
\begin{array}{cccc}
1 & \mathbf{2} & \mathbf{3} & 4 \\
\mathbf{5} & 6 & \mathbf{7} & 8 \\
9 & 10 & \mathbf{11} & 12 \\
\mathbf{13} & 14 & 15 & 16 \\
\mathbf{17} & 18 & \mathbf{19} & 20 \\
21 & 22 & \mathbf{23} & 24 \\
25 & 26 & 27 & 28 \\
\mathbf{29} & 39 & \mathbf{31} & 32 \\
33 & 34 & 35 & 36 \\
\mathbf{37} & 38 & 39 & 40 \\
\mathbf{41} & 42 & \mathbf{43} & 44 \\
45 & 46 & \mathbf{47} & 48 \\
49 & 50 & 51 & 52 \\
\mathbf{53} & 54 & 55 & 56 \\
57 & 58 & \mathbf{59} & 60 \\
\vdots & \vdots & \vdots & \vdots \\
\end{array}
$$

---

[5] Christian Goldbach (German: 1690–1764).

7.6 Counterexamples and Conjectures

We guide the reader through these data by asking some questions, arranged in order of increasing difficulty. (We have considered this form of exposition in Sect. 6.5.)

1. *Why is there just one prime in the second column and no primes at all in the fourth?*
2. *The first two rows combined contain four primes; are there other pairs of adjacent rows with the same property?*
3. *Does the table contain infinitely many primes?*
4. *There are rows without primes. Are there infinitely many of them?*
5. *Are there adjacent rows without primes? What about sequences of three or more adjacent rows without primes?*
6. *Are there infinitely many primes in the first column?*
7. *Are 'half' of the primes in column 1 and 'half' in column 3?*
8. *Are there infinitely many rows containing two primes?*

At the end of primary school, a pupil should be able to answer question 1, while in secondary school one could see why the answer to question 2 must be negative (one of three consecutive odd integers is divisible by 3). The answer to questions 3 to 7 is affirmative. Euclid's theorem (Sect. 7.5), which answers question 3, is normally taught in a first year university course, although we have seen that the proof does not require advanced ideas. Answering questions 4 and 5 does involve some ingenuity, but no major difficulty.

The mathematics needed to answer 6 and 7 becomes difficult. The proof that column 1 contains infinitely many primes, known since the 17th Century, is given today in an undergraduate course in number theory. The formulation of question 7 can be made precise, and the affirmative answer was given at the end of the 19th Century. Understanding the proof requires a solid background in analysis.

Many mathematicians would conjecture that the answer to question 8 is again positive, but at present nobody knows how to prove it. This is a variant of the twin primes conjecture stated above.

An accessible introduction to this beautiful part of mathematics is found in [38].

7.7 Wrong Arguments

In constructing a mathematical argument it's easy to make mistakes. In this section we identify some common faulty arguments: confusing examples with proofs, assuming what we are supposed to prove, mishandling functions. Other mistakes will be examined in Chap. 9 and in the exercises. Awareness of these problems should help us in avoiding them.

7.7.1 Examples Versus Proofs

The verification of a statement in specific cases does not constitute a form of proof. Our study of Euler's polynomial in Sect. 7.6 shows how misleading examples can be.

This state of affairs is peculiar to mathematics; in other scientific disciplines, such as biology, a proof of a statement consists of independent experimental verifications of its validity.

In our first example the fault is easy to spot.

Theorem. *For all primes p, the integer $2^p - 2$ is divisible by p.*

WRONG PROOF.
$$2^2 - 2 = 2 \cdot 1, \quad 2^3 - 2 = 3 \cdot 2, \quad 2^5 - 2 = 5 \cdot 6, \quad 2^7 - 2 = 7 \cdot 18, \quad \text{etc.} \quad \square$$

The theorem has been proved only for $p = 2, 3, 5, 7$.

The next example is similar, but not so clear-cut [36, p. 138f].

Theorem. *For all x, y and z, if $x + y < x + z$ then $y < z$.*

WRONG PROOF. Suppose $x + y < x + z$. Take $x = 0$. Then
$$y = 0 + y < 0 + z = z. \qquad \square$$

The mistake here is that we took x to be 0, which is a special value of x. This assumption is wholly unjustified, since the quantities x, y, z are controlled by an existential quantifier, and hence no condition may be imposed on them. By adding the assumption that $x = 0$, we have in fact proved the

WEAKER THEOREM. *For all y and z, if $0 + y < 0 + z$, then $y < z$.*

This is not what we claimed to prove.

7.7.2 Wrong Implications

Inappropriate handling of implications results in common mistakes. Instead of proving an implication, we may end up proving its converse; or we may assume the statement we are meant to prove, deduce from it a true statement, and believe we've completed the proof. These faulty deductions—of which we now show an example—are sometimes called circular arguments, or *non sequiturs*.[6]

Prove that $\sqrt{2} + \sqrt{6} < \sqrt{15}$

WRONG PROOF.
$$\sqrt{2} + \sqrt{6} < \sqrt{15} \Rightarrow (\sqrt{2} + \sqrt{6})^2 < 15$$
$$\Rightarrow 8 + 2\sqrt{12} < 15$$
$$\Rightarrow 2\sqrt{12} < 7$$
$$\Rightarrow 48 < 49. \qquad \square$$

---

[6] Latin for 'it does not follow'.

7.7 Wrong Arguments

We were supposed to prove P, where $P = (\sqrt{2} + \sqrt{6} < \sqrt{15})$. Instead we have assumed P, and correctly deduced from it the true statement $48 < 49$. However, the deduction $P \Rightarrow$ TRUE (unlike the deduction $P \Rightarrow$ FALSE) gives us no information about P, from Table (4.9). Indeed, had we started from the false statement $\sqrt{2} + \sqrt{6} < -\sqrt{15}$, we would have reached exactly the same conclusion.

There are two methods for fixing this problem.

First method: retracing the steps. We regard the chain of deductions displayed above as 'rough work'; then we start from the end and prove the chain of converse implications.

PROOF.

$$48 < 49 \Rightarrow \sqrt{48} < \sqrt{49}$$
$$\Rightarrow 2\sqrt{12} < 7$$
$$\Rightarrow 8 + 2\sqrt{12} < 15$$
$$\Rightarrow (\sqrt{2} + \sqrt{6})^2 < 15$$
$$\Rightarrow \sqrt{2} + \sqrt{6} < \sqrt{15}$$

where in the first and the last implications we have taken the positive square root of each side. We have proved the implication TRUE $\Rightarrow P$, from which we deduce that P is TRUE. □

Clearly, we could not have come up with such a proof without having done the 'rough work' first; the next method does not require this.

Second method: contradiction.

PROOF. Let us assume $\neg P$:

$$\sqrt{2} + \sqrt{6} \geqslant \sqrt{15} \Rightarrow (\sqrt{2} + \sqrt{6})^2 \geqslant 15$$
$$\Rightarrow 8 + 2\sqrt{12} \geqslant 15$$
$$\Rightarrow 2\sqrt{12} \geqslant 7$$
$$\Rightarrow 48 \geqslant 49.$$

We have proved that $\neg P \Rightarrow$ FALSE, which implies that $\neg P$ is FALSE, that is, P is TRUE. □

7.7.3 Mishandling Functions

Many things can go wrong when we deal with functions. For instance, we may apply a function to arguments outside its domain (e.g., forming the logarithm of zero), or invert a function incorrectly (e.g., taking the wrong sign of a square root).

We begin with a proof from calculus which isn't correct, even if it captures the essence of the argument. The chain rule for differentiating the composition of two functions states that if f and g are differentiable, then

$$(g \circ f)'(x) = g'(f(x))f'(x).$$

SLOPPY PROOF. Let ε be a real number. We compute

$$\frac{g(f(x+\varepsilon)) - g(f(x))}{\varepsilon} = \frac{g(f(x+\varepsilon)) - g(f(x))}{f(x+\varepsilon) - f(x)} \times \frac{f(x+\varepsilon) - f(x)}{\varepsilon}.$$

Letting ε tend to zero gives the desired result. □

The right-hand side is obtained by multiplying and dividing by $f(x+\varepsilon) - f(x)$. But this quantity may well be zero for non-zero ε (e.g., if f is constant), which invalidates the argument. A valid proof requires more delicacy.

PROOF. Let $y = f(x)$. Since g is differentiable at y, we write

$$g(y+\delta) - g(y) = g'(y)\delta + h(y, \delta)\delta$$

where h is a continuous function with $h(y, 0) = 0$. Specialising to $\delta = f(x+\varepsilon) - f(x)$, we find

$$\frac{g(f(x+\varepsilon)) - g(f(x))}{\varepsilon} = g'(f(x)) \frac{f(x+\varepsilon) - f(x)}{\varepsilon}$$
$$+ h(y, f(x+\varepsilon) - f(x)) \frac{f(x+\varepsilon) - f(x)}{\varepsilon}.$$

As ε tend to zero, the last term tends to zero, being the product of a function that tends to zero and a function that tends to a finite limit (because f is differentiable). This gives the desired result. □

Mishandling a non-invertible function may have unpredictable consequences:

WRONG THEOREM. *Every negative real number is positive.*

WRONG PROOF. Let α be a negative real number. Then

$$\alpha = -1|\alpha| = (-1)^1|\alpha| = (-1)^{2 \cdot \frac{1}{2}}|\alpha| = [(-1)^2]^{\frac{1}{2}}|\alpha|$$
$$= 1^{\frac{1}{2}}|\alpha| = |\alpha| > 0$$

as required. □

What's wrong with this proof? The function $x \mapsto x^2$ is not invertible, and to invert it we must restrict the domain. If we restrict it to the range $x \geqslant 0$ and take the positive sign for the square root, then the functions $x \mapsto x^2$ and $x \mapsto \sqrt{x}$ are inverse of each other: $x^{2 \cdot \frac{1}{2}} = \sqrt{x^2} = x$. However, if x is negative, these functions are inverse of each other only if we take the negative sign of the square root.

Thus the error occurred in the last equality

7.7 Wrong Arguments

$$1^{\frac{1}{2}} = 1$$

where, having squared a negative quantity, we mistakenly choose the positive sign of the square root. Exercise 7.3.2 deals with a more general instance of the same phenomenon.

7.8 Writing a Good Proof

Proofs come in all shapes and sizes. Correctness is, of course, imperative, but a good proof requires a lot more since the conflicting requirements of clarity and conciseness must be resolved. On the one hand, reading a proof is demanding, and any assistance will be welcome. On the other hand, a proof is a rigorous exercise in logic and the exposition must remain succinct and essential. The level of exposition and the amount of detail will depend on the mathematical maturity of the target audience.

There is one general rule, to be applied at all key junctures in a proof:

- SAY WHAT YOU PLAN TO DO.
- WHEN YOU'VE DONE IT, SAY SO.

We now examine statements of theorems and proofs that are less than optimal, and we seek to improve them. Our proofs are very detailed, and suitable for first-year university students. In some cases we also provide a concise version of the proof, aimed at expert readers.

Our first theorem is taken from a textbook.

BAD THEOREM. *If $x^2 \neq 0$, then $x^2 > 0$.*

BAD PROOF. If $x > 0$ then $x^2 = xx > 0$. If $x < 0$ then $-x > 0$, so $(-x)(-x) > 0$, i.e., $x^2 > 0$. □

The statement of the theorem is incomplete, in that there is no information about the quantity x. The theorem is an implication, but in the proof what has happened to the assumption $x^2 \neq 0$? If the assumption is obvious, one may leave it out, but one must judge whether the readers are sophisticated enough to see what it must be. Furthermore, this is a proof by cases (see Sect. 7.2), and it would be helpful if this were made explicit.

THEOREM. *For all real numbers x, if $x^2 \neq 0$, then $x^2 > 0$.*

PROOF. Let x be a real number such that $x^2 \neq 0$. Then $x \neq 0$, and we have two cases:

(i) $x < 0$. Then $-x > 0$, so $(-x)(-x) > 0$, that is, $x^2 > 0$.
(ii) $x > 0$. Then $xx = x^2 > 0$. □

The following statement establishes a simple relation between rational and irrational numbers.

BAD THEOREM. $\forall a, b \in \mathbb{R}, \ (a \in \mathbb{Q} \wedge b \notin \mathbb{Q}) \Rightarrow a + b \notin \mathbb{Q}$.

BAD PROOF: Suppose $a + b \in \mathbb{Q}$. Then $a + b = m/n$.
If $a \in \mathbb{Q}$, then $a = p/q$, and $b = m/n - p/q \in \mathbb{Q}$. □

The symbolic formulation of the theorem obscures its simple content. The proof is straightforward, but the use of unnecessary symbols (the numerator and denominator of the rational numbers serve no purpose) and the lack of comments make it less clear than it should be.

THEOREM. *The sum of a rational and an irrational number is irrational.*

PROOF. Let a and z be a rational and an irrational number, respectively. Consider the identity $z = (a + z) - a$. If $a + z$ were rational, then z, being the difference of two rational numbers, would also be rational, contrary to our assumption. Thus $a + z$ must be irrational. □

The choice of symbols keeps the rational and the irrational numbers well-separated in our mind.

Next we consider an arithmetical statement.

BAD THEOREM. n odd $\Rightarrow 8 | n^2 - 1$.

BAD PROOF.
n odd $\Rightarrow \exists j \in \mathbb{Z}, \ n = 2j + 1$;
$\therefore \ n^2 - 1 = 4j(j + 1)$;
$\forall j \in \mathbb{Z}, \ 2 \mid j(j + 1) \Rightarrow 8 \mid n^2 - 1$ □

This is a clumsy attempt to achieve conciseness via an entirely symbolic exposition. Combining words and symbols and adding some short explanations will improve readability and style. We shall see that a clearer proof need not be longer.

THEOREM. *The square of an odd integer is of the form* $8n + 1$.

DETAILED PROOF. Let k be an odd integer. We show that $k^2 = 8n + 1$, for some $n \in \mathbb{Z}$.

Since k is odd, we have $k = 2j + 1$, for some integer j. We find

$$k^2 = (2j + 1)^2 = 4j^2 + 4j + 1 = 4j(j + 1) + 1. \tag{1}$$

Now, one of j or $j + 1$ is even, and therefore their product is even. Thus we have $j(j + 1) = 2n$, for some n. Inserting this expression in (1) we obtain $k^2 = 8n + 1$, as desired. □

This proof is quite detailed, and can safely be shortened.

PROOF. An odd integer k is of the form $k = 2j + 1$, for some integer j. A straightforward manipulation gives $k^2 = 4j(j + 1) + 1$. Our claim now follows from the observation that the product $j(j + 1)$ is necessarily even. □

The adverb 'necessarily' signals that the conclusion (the product $j(j+1)$ is even) requires a moment's thought. The expression 'straightforward manipulation' is used

7.8 Writing a Good Proof

to omit some steps in the derivation, while warning the reader that some calculations are required. In such a circumstance, the adjective 'straightforward' is preferable to 'easy', which is subjective, or 'trivial', which carries a hint of arrogance. So it's appropriate to claim that the arithmetical identity

$$29 = \left(2 + \frac{\sqrt{-1}(1+\sqrt{5})}{2}\right)\left(2 - \frac{\sqrt{-1}(1+\sqrt{5})}{2}\right)$$
$$\times \left(2 + \frac{\sqrt{-1}(1-\sqrt{5})}{2}\right)\left(2 - \frac{\sqrt{-1}(1-\sqrt{5})}{2}\right)$$

can be verified with a 'straightforward calculation'.

In the following example everything needs re-writing: definition, theorem, and proof.

BAD DEFINITION. Let q_j be digits. We define

$$\overline{q_1 \cdots q_n} = q_1 \cdots q_n q_1 \cdots$$

BAD THEOREM. *Let* $x = m + 0.q'_1 \cdots q'_k \overline{q_1 \cdots q_n}$, *where* $m \in \mathbb{Z}$. *Then* $x \in \mathbb{Q}$.

BAD PROOF: Let $A = 0.q'_1 \cdots q'_k$, $B = 0.q_1 \cdots q_n$. Then $A, B \in \mathbb{Q}$, and

$$0.\overline{q_1 \cdots q_n} = B\left(1 + \frac{1}{10^n} + \frac{1}{10^{2n}} + \cdots\right) = B\frac{10^n}{10^n - 1} =: C \in \mathbb{Q}.$$

Thus $x = m + A + C10^{-k} \in \mathbb{Q}$. □

In the definition, the decimal nature of the digits is not mentioned and the periodicity of the digit sequence could be made more evident. In the statement of the theorem the number x is represented with an inappropriate hybrid notation, as the sum of a decimal number and a number expressed symbolically. The proof lacks explanations and the choice of symbols is not ideal.

As we did before, we state the theorem with words, which is more effective. We introduce the notation *after* the statement of the theorem but *before* the beginning of the proof. This arrangement should be considered if the notation is needed outside the confines of the proof, or if the proof is heavy and needs lightening up, or simply as a variation from the rigid definition-theorem-proof format. We number the various steps in the argument for future reference.

THEOREM. *Every number whose decimal digits eventually repeat is rational.*

1. Before proving the theorem, we establish some notation. We denote a string (finite or infinite) of decimal digits by $d_1 d_2 \cdots$, with $d_k \in \{0, 1, \ldots, 9\}$; the over-bar denotes a pattern of digits which repeats indefinitely:

$$\overline{d_1 \cdots d_n} = d_1 \cdots d_n d_1 \cdots d_n \cdots. \tag{7.9}$$

DETAILED PROOF.
2. Let x be a real number with eventually repeating decimal digits. Then x has a decimal representation of the type

$$x = d'_0 \cdots d'_j . d'_{j+1} \cdots d'_m \overline{d_1 \cdots d_n}$$

for some $m \geqslant 0$, $0 \leqslant j \leqslant m$, and $n \geqslant 1$. The primed digits form the aperiodic part of the digit sequence.

3. We shall compute the value of x in terms of two integers M and N, given by:

$$M = d'_0 \cdots d'_m = \sum_{k=0}^{m} d'_k 10^{m-k} \qquad N = d_1 \cdots d_n = \sum_{k=1}^{n} d_k 10^{n-k}.$$

First, we get rid of the aperiodic part; we shift it to the left of the decimal point by multiplying by a suitable power of 10, and then we subtract it off, to get

$$10^{m-j} x - M = 0.\overline{d_1 \cdots d_n}. \tag{7.10}$$

4. Next we compute, using (7.9)

$$0.\overline{d_1 \cdots d_n} = \frac{N}{10^n} + \frac{N}{10^{2n}} + \frac{N}{10^{3n}} + \cdots$$

$$= N \left(\frac{1}{10^n} + \frac{1}{10^{2n}} + \frac{1}{10^{3n}} + \cdots \right)$$

$$= N \sum_{k=1}^{\infty} \left(\frac{1}{10^n} \right)^k = \frac{N}{10^n - 1}.$$

5. In the last step we have used the formula of the sum of the geometric series

$$\sum_{k=1}^{\infty} q^k = \frac{q}{1-q} \qquad |q| < 1.$$

From Eq. (7.10) and the result above, we obtain

$$10^{m-j} x - M = \frac{N}{10^n - 1}$$

and hence

$$x = 10^{j-m} \left(M + \frac{N}{10^n - 1} \right).$$

The right-hand side of this equation consists of sums and products of rational numbers, and is therefore rational. □

7.8 Writing a Good Proof

Let us examine the main steps in the proof:

1. Since we don't begin the proof straight away, we say so. The key notation (the over-bar) is established first in words, then in symbols. The periodicity of the digit sequence is clearer.
2. In the theorem there is a hidden quantifier, and the proof begins accordingly. Then we introduce the notation for x, and we say why.
3. At every opportunity, we declare our intentions. A coherent choice of symbols improves readability.
4. In this passage, one must decide what constitutes an appropriate amount of details: we have been cautious here.
5. A reminder of a well-known summation formula.

We give a concise version of the same proof.

PROOF. Let x be a real number with eventually repeating decimal digits d_k. Without loss of generality, we may assume that the integer part of x is zero, and that the fractional part is purely periodic:

$$x = 0.\overline{d_1 \cdots d_n}. \qquad (1)$$

(Any real number may be reduced to this form by first multiplying by a power of 10, and then subtracting an integer, and neither operation affects the tail of the digit sequence.) Defining the decimal integer $M = d_1 \cdots d_n$, we find, from (1)

$$x = \frac{M}{10^n} + \frac{M}{10^{2n}} + \frac{M}{10^{3n}} + \cdots = M \sum_{k=1}^{\infty} \left(\frac{1}{10^n}\right)^k = \frac{M}{10^n - 1}. \qquad (7.11)$$

We see that x is rational. □

The expression 'without loss of generality' indicates that the restriction being introduced does not weaken the argument in any way (see also Sect. 5.2). A brief parenthetical remark is added for clarification: it too could be omitted.

Exercise 7.1 You are given cryptic proofs of mathematical statements. Rewrite them in a good style, with plenty of explanations.

1. *Prove that the line through the point $(4, 5, 1) \in \mathbb{R}^3$ parallel to the vector $(1, 1, 1)^T$ and the line through the point $(5, -4, 0)$ parallel to the vector $(2, -3, 1)^T$ intersect at the point $(1, 2, -2)$. (The symbol T denotes transposition.)*
   BAD PROOF.
   $(4, 5, 1)^T + \lambda (1, 1, 1)^T = (4 + \lambda, 5 + \lambda, 1 + \lambda)^T;$
   $= (5, -4, 0)^T + \mu (2, -3, 1)^T = (5 + 2\mu, -4 - 3\mu, \mu)^T.$
   $\therefore 4 + \lambda = 5 + 2\mu, \ 5 + \lambda = -4 - 3\mu, \ 1 + \lambda = \mu.$
   $\Rightarrow 4 + \lambda = 5 + 2(1 + \lambda) \Rightarrow \lambda = -3, \ \mu = -2.$
   $\therefore \mathbf{v} = (1, 2, -2)^T.$ □

2. *Prove that the real function* $x \mapsto 3x^4 + 4x^3 + 6x^2 + 1$ *is positive.*
   BAD PROOF.

   $f'(x) = 12x(x^2 + x + 1) = 0 \Leftrightarrow x = 0;$
   $f''(0) > 0, f(0) > 0 \Rightarrow$ positive. □

3. *Prove that for all real values of* a, *the line*

   $$y = ax - \left(\frac{a-1}{2}\right)^2$$

   is tangent to the parabola $y = x^2 + x$.
   BAD PROOF.

   $ax - (a-1)^2/4 = x^2 + x;$
   $\Rightarrow x^2 + x(1-a) + (a-1)^2/4 = 0;$
   $\Rightarrow x = (a-1)/2.$
   For this x, dy/dx is the same. □

4. *Prove that the vector* $(1, 0, 1)$ *does not belong to the subspace of* $\mathbb{R}^3$ *generated by the vectors* $(3, 1, 2)$ *and* $(-2, 1, 0)$.
   BAD PROOF:

   $3t - 2s = 1; t + s = 0; 2t = 1.$
   $\Rightarrow t = 1/2 \Rightarrow s = -1/2 \Rightarrow 3/2 = 0,$ contradiction. □

Exercise 7.2 The following text has several faults: (a) explain what they are; (b) write an appropriate revision.
WRONG THEOREM. *For all numbers* x *and* y,

$$\frac{x^2 + y^2}{|xy|} > 2.$$

WRONG PROOF. For

$$\frac{x^2 + y^2}{|xy|} > 2 \Rightarrow x^2 + y^2 > 2xy$$
$$\Rightarrow x^2 - 2xy + y^2 > 0$$
$$\Rightarrow (x - y)^2 > 0.$$

The last equation is trivially true, which proves it. □

Exercise 7.3 Identify the flaw in the proof of the following statements.

1. *The integral of a positive function may be negative.*
   WRONG PROOF. The function $f(x) = 1/x^2$ is positive. We compute

   $$I(a, b) = \int_a^b \frac{dx}{x^2} = \left[-\frac{1}{x}\right]_a^b = \frac{1}{a} - \frac{1}{b}.$$

7.8 Writing a Good Proof

We find that $I(-1, 1) = -2 < 0$. □

2. The complex exponential function is constant.
   WRONG PROOF. Let θ be an arbitrary complex number. We find
   $$e^{i\theta} = e^{i\theta \frac{2\pi}{2\pi}} = e^{2\pi i \frac{\theta}{2\pi}} = \left(e^{2\pi i}\right)^{\frac{\theta}{2\pi}} = 1^{\frac{\theta}{2\pi}} = 1.$$
   □

Exercise 7.4 Write the first few sentences of the proof of each statement, introducing all relevant notation in an appropriate order, and identifying the RTP. (For this task, understanding the meaning of the statements is unimportant.)

1. On a compact set every continuous function is uniformly continuous.
2. Every valuation of a field with prime characteristic is non-archimedean.
3. No polynomial with integer coefficients, not a constant, can be prime for all integer values of the indeterminate.
4. Any hyperbolic matrix in $SL_2(\mathbb{Z})$ is conjugate to a matrix with non-negative coefficients.
5. The only bi-unique analytic mapping of the interior of the unit disc into itself is a linear fractional transformation.
6. In a field with a non-archimedean absolute value, the connected component of any point is the set consisting of only that point.
7. The characteristic polynomial of the incidence matrix of a Pisot substitution is irreducible over $\mathbb{Q}$.
8. The first Betti number of a compact and orientable Riemannian manifold of positive Ricci curvature is zero.
9. Every differential of the second or third kind differs from some normal differential by a differential of the first kind.
10. A real function continuous in a closed interval attains all values between its maximum and minimum.
11. A subset of a metric space is open if and only if its complement is closed.
12. A stable set of a graph is maximum if and only if there exists no maximal alternating sequence of odd length.
13. A bi-infinite sequence over the alphabet $\{0, 1\}$ is Sturmian if and only if it is balanced and not eventually periodic.
14. The curvature of a hermitian manifold vanishes if and only if it is possible to choose a parallel field of orthonormal holomorphic frames in a neighbourhood of each point of the manifold.

Exercise 7.5 You are asked to help the reader approach the following problem:

Prove that for all integers m, n, we have $m\mathbb{Z} + n\mathbb{Z} = \gcd(m, n)\mathbb{Z}$.

Formulate a number of easier sub-problems of increasing difficulty, eventually leading to the original problem.

Chapter 8
Induction

Induction is a fundamental method of proof, used in every corner of mathematics. It is applicable to statements of the form

$$\forall n \in \mathbb{N}, \quad \mathscr{P}(n) \qquad (8.1)$$

where $\mathscr{P}$ is a predicate over $\mathbb{N}$. This sentence says that all natural numbers have property $\mathscr{P}$.

Induction is first met in the proof of summation identities and inequalities:

1	$\sum_{k=0}^{n} x^k = \dfrac{1 - x^{n+1}}{1 - x}$	$x \neq 1$	Sum of geometric progression
2	$(x + y)^n = \sum_{k=0}^{n} \binom{n}{k} x^k y^{n-k}$		Binomial theorem
3	$(1 + x)^n \geqslant 1 + nx$	$x > -1$	Bernoulli inequality[1]
4	$\prod_{k=1}^{n} x_k \leqslant \left(\dfrac{1}{n} \sum_{k=1}^{n} x_k\right)^n$	$x_k > 0$	Arithmetico-geometric mean (AGM) inequality
5	$\left\lvert \sum_{k=1}^{n} x_k y_k \right\rvert^2 \leqslant \sum_{k=1}^{n} \lvert x_k \rvert^2 \sum_{k=1}^{n} \lvert y_k \rvert^2$		Cauchy-Schwarz inequality[2]

(The variables are complex numbers in 1, 2, and real numbers in 3, 4, 5.)

The expression (8.1) unfolds into an infinite sequence of boolean expressions:

$$\mathscr{P}(1), \; \mathscr{P}(2), \; \mathscr{P}(3), \; \ldots$$

---

[1] Jacob Bernoulli (Swiss: 1654–1705).
[2] Augustin-Louis Cauchy (French: 1789–1857); Hermann Schwarz (German: 1843–1921).

© Springer-Verlag London 2014
F. Vivaldi, *Mathematical Writing*, Springer Undergraduate Mathematics Series,
DOI 10.1007/978-1-4471-6527-9_8

and to prove it we must show that all expressions in the sequence are true:

$$T, T, T, \ldots.$$

For instance, the Bernoulli inequality unfolds as follows:

$$1 + x \geqslant 1 + x, \quad (1+x)^2 \geqslant 1 + 2x, \quad (1+x)^3 \geqslant 1 + 3x, \quad \ldots$$

each inequality being valid for all real numbers $x > -1$.

A special form of mathematical argument is needed for this purpose, called **mathematical induction**. It takes several forms, and we will examine four:

- (A) The well-ordering principle.
- (B) The infinite descent method.
- (C) The induction principle.
- (D) The strong induction principle.

We will show that these principles are equivalent in the sense that any one of them implies all the others. The different forms of the principle are appropriate for different situations.

The term mathematical induction was coined by De Morgan around 1840. The attribute 'mathematical' is used to differentiate this concept from the homonymous concept in philosophy, which has a different meaning.[3] The induction principle—in any of the above forms—is one of the five **Peano axioms**[4] which characterise the set $\mathbb{N}$ of natural numbers.

8.1 The Well-Ordering Principle

The **well-ordering** (or **least counter example**) **principle** states that

(A) *Every non-empty set of natural numbers contains a least element.*

This principle is clearly false for subsets of $\mathbb{Z}$, $\mathbb{Q}$ or $\mathbb{R}$, and, less trivially so, for subsets of $\mathbb{Q}^+$ or $\mathbb{R}^+$, or of any real or rational interval. A possible strategy for proving (8.1) is to combine well-ordering with contradiction. Such a proof will begin as follows:

---

[3] In philosophy, an inductive argument is one that uses very strong premises to support the *probable* truth of a conclusion.

[4] Giuseppe Peano (Italian: 1858–1932).

8.1 The Well-Ordering Principle

PROOF. Suppose (8.1) is false. Let k be the least natural number n for which $\mathscr{P}(n)$ is false. ...

Then with this k we try to deduce a contradiction, using the knowledge that $\mathscr{P}(n)$ is true for $n = 1, \ldots, k-1$. The integer k defined in the proof exists by virtue of the well-ordering principle: it is the least element of the set $\{n \in \mathbb{N} : \neg \mathscr{P}(n)\}$, which is a subset of $\mathbb{N}$. By assumption, this subset is non-empty.

Let us apply this strategy to some specific problems.

Proposition. *If $n \geqslant 7$, then $n! > 3^n$.*

PROOF. Suppose this statement is false, and let k be the least integer $n \geqslant 7$ such that $n! \leqslant 3^n$. We know that $k > 7$, since

$$7! = 5040 > 2187 = 3^7.$$

Put $j = k - 1$. Then $j \geqslant 7$, and hence $j! > 3^j$ by choice of k. So

$$k! = (j+1)j! > (j+1)3^j > 3 \cdot 3^j = 3^k$$

contradicting the choice of k. □

The initial verification that $k > 7$ creates enough room for the principle to work. We note that $6! = 720$ and $3^6 = 729$, so the inequality $n \geqslant 7$ is the best possible one. Under these circumstances, we say that $n \geqslant 7$ is a **strict bound**. Our next statement is more substantial:

Theorem. *Every integer greater than 1 is a product of one or more prime numbers.*

PROOF. Suppose the statement is false, and let k be the smallest integer greater than 1 which is not a product of primes. Then k is not prime, so $k = mn$ for two smaller integers $m, n \geqslant 2$. Since m, n are smaller than k and greater than 1, they are products of prime numbers. So $k = mn$ is also a product of primes, which contradicts the choice of k. □

This form of mathematical induction can be very effective. The assumption that k is the smallest counterexample puts the maximum amount of information on the table; it gives us the feeling that we are examining a concrete object.

8.2 The Infinite Descent Method

This is a second form of mathematical induction, developed in the 17th Century by Fermat.[5] It exploits the following variant of the well-ordering principle:

(B) *There is no infinite decreasing sequence of natural numbers.*

---

[5] Pierre de Fermat (French: 1601 or 1607/08–1665).

This principle follows from the well-ordering principle $((A) \Rightarrow (B))$: if there were an infinite decreasing sequence of natural numbers

$$n_1 > n_2 > n_3 > \cdots$$

then the set $\{n_1, n_2, n_3, \ldots\}$, which is a non-empty subset of $\mathbb{N}$, would have no smallest element. Conversely, well-ordering follows from descent $((B) \Rightarrow (A))$: if there were a non-empty set $S \subset \mathbb{N}$ without a smallest element, then choosing an arbitrary element $n_1 \in S$, we could find $n_2 \in S$ with $n_2 < n_1$. Repeating this process we would generate an infinite decreasing sequence of natural numbers.

Fermat applied this method extensively to arithmetical problems. He pointed out that infinite descent is a method for proving that certain properties or relations for whole numbers are *impossible*, and for this purpose he combined descent with contradiction—as we did for well-ordering. To prove that no positive integer can have a certain set of properties, we assume that a positive integer does have these properties, and we try to deduce that there is a smaller positive integer with the same set of properties. If we succeed, then, by the same argument, these properties would hold for some smaller positive integer, and so forth *ad infinitum*. Because a sequence of natural numbers cannot decrease indefinitely, no positive integer can have the stated properties.

Infinite descent was brought to prominence by Euler, who proved with it that it is impossible to find natural numbers x, y, z such that

$$x^3 + y^3 = z^3.$$

This is a special case of the celebrated **Fermat's last theorem**, in which the exponent 3 is replaced by an arbitrary integer $n > 2$. In keeping with this tradition, we now apply the method of infinite descent to prove the non-existence of integer solutions of a polynomial equation.

Theorem. *For every prime p the equation $px^2 = y^2$ has no solutions in the natural numbers.*

This statement is equivalent to the irrationality of $\sqrt{p}$, as one sees by dividing each term of the equation by x^2, and then taking the positive square root. The argument presented here is a variant and a generalisation of that given in Sect. 7.1 for the irrationality of $\sqrt{2}$.

PROOF. Suppose that there are natural numbers n_0, n_1, such that $pn_1^2 = n_0^2$. So pn_1^2 is the square of a natural number. Since p divides $n_0^2 = n_0 n_0$, by the basic property of prime numbers p must divide n_0, and we have $n_0 = pn_2$ for some natural number n_2. Then $pn_1^2 = p^2 n_2^2$, and hence $n_1^2 = pn_2^2$. So $n_2 < n_1$ and pn_2^2 is the square of a natural number.

Repeating the argument, there is $n_3 < n_2$ such that pn_3^2 is the square of a natural number. This continues for ever, which is impossible. Hence the equation $px^2 = y^2$ has no solution, as stated. □

8.3 Peano's Induction Principle

This is a third form of mathematical induction; it was proposed by Dedekind[6] and formalised by Peano. The **principle of induction** states that:

(C) If $\mathscr{P}$ is a predicate over $\mathbb{N}$ such that

  i) $\mathscr{P}(1)$ is true;   ii) $\forall k \in \mathbb{N}, \quad \mathscr{P}(k) \Rightarrow \mathscr{P}(k+1)$;

  then $\mathscr{P}(n)$ is true for all $n \in \mathbb{N}$.

The first condition is called the **basis** of the induction (or the **base case**); the second the **inductive step**. The bi-unique correspondence between predicates $\mathscr{P}$ over $\mathbb{N}$ and subsets S of $\mathbb{N}$, given by [see Eqs. (4.15 and 4.16)]

$$S = \{n \in \mathbb{N} : \mathscr{P}_S(n)\}, \qquad \mathscr{P}_S(x) = (x \in S)$$

allows us to reformulate the induction principle in the language of sets.

(C') If S is a subset of $\mathbb{N}$ such that

  i) $1 \in S$;   ii) $\forall k \in \mathbb{N}, \quad k \in S \Rightarrow (k+1) \in S$;

  then $S = \mathbb{N}$.

The principle of induction follows from well-ordering $((A) \Rightarrow (C))$. To show this, we define the set

$$S' = \mathbb{N} \setminus S = \{n \in \mathbb{N} : \neg \mathscr{P}(n)\}.$$

We prove that S' is empty by contradiction. If S' is non-empty, then, by the well-ordering axiom, S' has a least element k. Since $1 \in S$, we have $k \neq 1$, and since k is the smallest element of S', it follows that $k - 1 \in S$. Then, putting $j = k - 1$, we find from condition ii) that $j + 1 = k \in S$, which contradicts the fact that $k \in S'$.

The base case of Peano's induction could be any integer, not just 1. Indeed if $\mathscr{P}(n)$ is valid for all $n \geq k$, then by letting $m = n - k + 1$, the range $n \geq k$ becomes $m \geq 1$.

Peano's induction provides straightforward proofs of finite sums and products formulae:

$$\forall n \in \mathbb{N}, \quad \sum_{j=1}^{n} a_j = F(n) \qquad (8.2)$$

$$\forall n \in \mathbb{N}, \quad \prod_{j=1}^{n} a_j = G(n) \qquad (8.3)$$

---

[6] Richard Dedekind (German: 1831–1916).

where (a_j) is a sequence of numbers (or more generally of elements of a commutative ring), and F and G are explicit functions of n, hopefully easier to compute than the original sum or product.

In an inductive proof of (8.2), the base case consists of verifying that $a_1 = F(1)$. For the inductive step, we assume that the formula holds for some $n = k$, and using the induction hypothesis we obtain:

$$\sum_{j=1}^{k+1} a_j = \sum_{j=1}^{k} a_j + a_{k+1} = F(k) + a_{k+1}.$$

Thus the proof reduces to the verification of the statements

$$a_1 = F(1), \qquad F(k) + a_{k+1} = F(k+1) \quad k \geqslant 1 \tag{8.4}$$

which no longer involve summation. For example, to prove the formula

$$\sum_{j=1}^{n} j^2 = \frac{n(n+1)(2n+1)}{6} \qquad n = 1, 2, \ldots$$

we must verify that

$$1^1 = \frac{1 \cdot 2 \cdot 3}{6}, \qquad \frac{k(k+1)(2k+1)}{6} + (k+1)^2 = \frac{(k+1)(k+2)(2k+3)}{6}.$$

For a product formula, the expressions (8.4) are replaced by

$$a_1 = G(1), \qquad G(k) \times a_{k+1} = G(k+1) \quad k \geqslant 1. \tag{8.5}$$

In some cases the identities (8.4) and (8.5) may be established using a computer algebra system. This is an instance of **computer-assisted proof**[7]—see Exercise 8.1.

A similar arrangement applies to the inductive proof of inequalities. However, even in the simplest cases, such proofs are not as mechanical as those of summation and product formulae, as the following example illustrates.

Proposition. *If $n > 3$, then $2^n \geqslant n^2$.*

PROOF. We prove it by induction on n. The base case $n = 4$ is immediate: $2^4 \geqslant 4^2$. Assume now that for some $k \geqslant 4$ we have $2^k \geqslant k^2$. Then

$$2^{k+1} = 2 \cdot 2^k \geqslant 2 \cdot k^2$$

where the inequality follows from the induction hypothesis. To complete the proof, we must show that for all $k \geqslant 4$ we have $2 \cdot k^2 \geqslant (k+1)^2$. Now the polynomial

---

[7] This concept raises controversy among mathematicians.

8.3 Peano's Induction Principle

$$p(k) = 2 \cdot k^2 - (k+1)^2 = k^2 - 2k - 1$$

has roots $1 \pm \sqrt{2}$, with

$$-1 < 1 - \sqrt{2} < 1 + \sqrt{2} < 3.$$

Thus, for $k \geqslant 3$, we have $p(k) > 0$, or $2k^2 > (k+1)^2$. Hence $2^{k+1} \geqslant (k+1)^2$, completing the induction. □

The next example warns us that a straightforward inductive strategy doesn't always work.

Proposition. *For all integers $n \geqslant 0$ and real numbers $x > -1$, we have*

$$(1+x)^n > nx. \tag{8.6}$$

We attempt to prove this by induction on n.

Base case: $n = 0$. Since $x > -1$, we have $(1+x)^0 = 1 > 0 = 0x$, as desired.
Inductive step: Assume (8.6) for some $n = k \geqslant 0$. Then

$$\begin{aligned}(1+x)^{k+1} &= (1+x)(1+x)^k \\ &> (1+x)kx \quad \text{(by induction hypothesis)} \\ &= kx + kx^2 \quad ???\end{aligned}$$

We don't seem to be getting anywhere.

PROOF. We prove the *stronger* inequality $(1+x)^n \geqslant 1 + nx$. This is the **Bernoulli inequality**—see p. 139.
Base case: $n = 0$. Since $x > -1$, we have $(1+x)^0 = 1 \geqslant 1 + 0x$.
Inductive step: Assume the inequality for $n = k \geqslant 0$. Then

$$\begin{aligned}(1+x)^{k+1} &= (1+x)(1+x)^k \\ &\geqslant (1+x)(1+kx) \quad \text{(by induction hypothesis)} \\ &= 1 + (k+1)x + kx^2 \geqslant 1 + (k+1)x.\end{aligned}$$

The inductive step is complete. □

8.4 Strong Induction

In an induction proof, the successful completion of the inductive step rests on the assumption that to prove $\mathscr{P}(k+1)$ we only need $\mathscr{P}(k)$. Should we instead (or in addition) need $\mathscr{P}(k-1)$, then this method would collapse.

This situation is not unusual. Suppose we want to prove by induction that every integer is a product of prime numbers. To prove that this statement is true for, say,

$k+1 = 12$, the preceding case $k = 11$ is of no use. We would need instead $k = 3$ and $k = 4$.

The rescue is a form of induction called **strong induction** (or **complete induction**), which is formulated as follows:

(D) *If $\mathscr{P}$ is a predicate over $\mathbb{N}$ such that*

  i) *$\mathscr{P}(1)$ is true;*
  ii) *for any $k \in \mathbb{N}$, if $\mathscr{P}(1)$, $\mathscr{P}(2)$, ..., $\mathscr{P}(k)$ are true, so is $\mathscr{P}(k+1)$;*

  then $\mathscr{P}(n)$ is true for all $n \in \mathbb{N}$.

As with Peano's induction, the base case of strong induction can be any integer, and there is an equivalent formulation in terms of sets:

(D′) *If S is a subset of $\mathbb{N}$ such that*

  i) $1 \in S$;
  ii) *for any $k \in \mathbb{N}$, $\{1, 2, \ldots, k\} \subset S \Rightarrow k+1 \in S$;*

  then $S = \mathbb{N}$.

Strong induction follows from Peano's induction ($(C) \Rightarrow (D)$), since the latter has fewer conditions. (One could say that induction is *stronger* than strong induction!)

In Sect. 8.1 we proved that every integer greater than 1 is a product of one or more primes by combining well-ordering with contradiction; we now prove the same statement using strong induction. The two proofs employ two variants of the same argument, giving a tangible illustration of the equivalence of well-ordering and strong induction.

PROOF. We use strong induction. The integer 2 is prime, which establishes the base case. Suppose that for some $k > 1$ every integer n with $1 < n \leqslant k$ is a product of primes, and consider $k+1$. If $k+1$ is prime, then there's nothing to prove. If $k+1$ is not a prime then there are integers m, n such that $mn = k+1$ and $1 < m, n \leqslant k$. By the induction hypothesis, m and n are products of primes, and hence so is $k+1$. We have completed the inductive step. □

With the help of strong induction, Euclid's elegant proof of the infinitude of primes (Sect. 7.5) yields an **upper bound** (or an **estimate**) for the nth prime number.

Theorem 8.1 *For all $n \geqslant 1$, the nth prime p_n satisfies the upper bound $p_n \leqslant 2^{2^{n-1}}$.*

PROOF. We use strong induction. The base case is clear: $p_1 = 2 \leqslant 2^{2^0} = 2$. Assuming that for some $k \geqslant 1$ we have $p_j \leqslant 2^{2^{j-1}}$ for $j = 1, 2, \ldots k$, we have

8.4 Strong Induction

$$p_{k+1} \leq 1 + \prod_{j=1}^{k} p_j \leq 1 + \prod_{j=1}^{k} 2^{2^{j-1}}$$

$$= 1 + 2^{\sum_{j=1}^{k} 2^{j-1}} = 1 + 2^{2^k - 1} < 2^{2^k}$$

where the first inequality follows from Euclid's argument (the right-hand side is divisible by a prime greater than p_k, see (7.7)), while the second inequality follows from the induction hypothesis. The proof is complete. $\square$

To complete the proof that the four formulations of the induction principle are equivalent, it remains to show that well-ordering follows from strong induction $((D) \Rightarrow (A))$. When this is done, we will have proved that

$$(A) \Leftrightarrow (B) \quad \text{and} \quad (A) \Rightarrow (C) \Rightarrow (D) \Rightarrow (A)$$

and equivalence follows from a loop of implications (cf. Sect. 7.4.1).

We use the both ends method. Assume that strong induction holds but that there is a non-empty subset S of $\mathbb{N}$ without a least element. Let $S' = \mathbb{N} \smallsetminus S$. Then $1 \in S'$, for otherwise 1 would be the least element of S. Likewise, if for some $k \geq 1$ we have $\{1, \ldots, k\} \subset S'$, then we must also have $k+1 \in S'$, because otherwise $k+1$ would be the smallest element of S. Hence, from strong induction, we have that $S' = \mathbb{N}$, and hence $S = \emptyset$, contradicting the hypothesis $S \neq \emptyset$.

8.5 Good Manners with Induction Proofs

Induction proofs are easy to structure, as there is a clear procedure to follow. But even if the steps in a proof are predictable, they should be spelled out clearly.

1. At the beginning of the proof, say with respect to what variable induction is performed, unless this is obvious.

   We proceed by induction on the degree d of the polynomial.

2. If the induction variable is n, then in the inductive step it is considered good practice to go from $n = k$ to $n = k+1$ rather than from n to $n+1$.

3. In the inductive step, say at what point the induction hypothesis is used.

   …where the last inequality follows from the induction hypothesis.

4. Say when the inductive step is complete.

   This completes the induction.

If an induction argument is easy, then it may be tempting to skip it altogether.

Let $x_1 = 2$, and let $x_{n+1} = x_n^2$, for $n \geq 1$. Then $x_n = 2^{2^{n-1}}$.

The last sentence is a bit blunt, and it could be replaced by more considerate sentences, such as:

A straightforward induction argument shows that $x_n = 2^{2^{n-1}}$.

Then $x_2 = 2^2$, $x_3 = 2^{2^2}$, *and, in general,* $x_n = 2^{2^{n-1}}$.

Exercise 8.1 Prove the following formulae by induction, first analytically, then with a computer-assisted proof.

1. $\sum_{k=1}^{n}(2k-1) = n^2$

2. $\sum_{k=1}^{n} k! \cdot k = (n+1)! - 1$

3. $\sum_{k=1}^{n} k^3 = \left(\sum_{k=1}^{n} k\right)^2$

4. $\sum_{k=1}^{n} \frac{1}{k(k+1)} = \frac{n}{n+1}$

5. $\sum_{k=2}^{n} \frac{1}{k^2-1} = \frac{3}{4} - \frac{2n+1}{2n(n+1)}$

6. $\prod_{k=0}^{n-1}(1+x^{2^k}) = \frac{1-x^{2^n}}{1-x}$ $\quad x \neq 1$.

Exercise 8.2 Prove by induction all identities and inequalities listed at the beginning of this chapter. (The Bernoulli inequality is proved in Sect. 8.3.)

Exercise 8.3 Prove the following statements by induction.

1. The sum of the cubes of three consecutive natural numbers is a multiple of 9.
2. The sum of the internal angles of a polygon with n sides is $(n-2)\pi$.
3. The negation of a boolean expression with n leading quantifiers is obtained by interchanging quantifiers and negating the predicate.
4. The regions in which the plane is subdivided by n lines may be coloured with just two colours in such a way that no two neighbouring regions have same colour.

Exercise 8.4 Find each formula, and then prove it by induction.

1. The formula for the cardinality of the power set of a finite set.
2. The formula for the higher-order derivatives of the product of two functions.
3. The formula for the derivative of the repeated composition of a function with itself.
4. The formula for the maximum number of regions defined by n lines in the plane.

8.5 Good Manners with Induction Proofs

Exercise 8.5 Identify the flaw in the proof of the following false statements.

1. WRONG THEOREM. *For any natural number n, the following holds:*

$$2 + 4 + \cdots + 2n = (n-1)(n+2).$$

WRONG PROOF. Assume that $2 + 4 + \cdots + 2k = (k-1)(k+2)$ for some $k \in \mathbb{N}$. Then

$$\begin{aligned}
2 + 4 + \cdots + 2(k+1) &= (2 + 4 + \cdots + 2k) + 2(k+1) \\
&= (k-1)(k+2) + 2(k+1) \quad \text{(by the induction hypothesis)} \\
&= k(k+3) = ((k+1) - 1)((k+1) + 2)
\end{aligned}$$

which completes the induction. □

2. WRONG THEOREM. *For any $n \in \mathbb{N}$, if $\max(i, j) = n$ for two natural numbers i and j, then $i = j$. Thus any two natural numbers are equal.*

WRONG PROOF. The statement is true for $n = 1$: if $\max(i, j) = 1$, then, necessarily, $i = j = 1$. Now assume the statement to be true for some $k \geq 1$, and let i, j be natural numbers such that $\max(i, j) = k + 1$. Then $\max(i - 1, j - 1) = k$, and by the induction hypothesis, we have that $i - 1 = j - 1$. But then $i = j$, and therefore our statement is true for $n = k + 1$. This completes the induction. □

3. WRONG THEOREM. (Pólya[8]) *In any group of n horses, all horses have the same colour.*

WRONG PROOF. The statement is clearly true for $n = 1$. Assume now that the statement is true for some $n = k \geq 1$, and consider an arbitrary collection of $k + 1$ horses.

$$\underbrace{\bullet\bullet \;\cdots\; \bullet\bullet}_{k+1}$$

By the induction hypothesis, the first k horses have the same colour, and so do the last k horses.

$$\overbrace{\underbrace{\bullet\bullet \;\cdots\; \bullet}\bullet}^{k}_{k}$$

But then all $k + 1$ horses have the same colour as the $k - 1$ horses common to the two sets.

$$\bullet\underbrace{\bullet \;\cdots\; \bullet}_{k-1}\bullet$$

This completes the inductive step. □

---

[8] George Pólya (Hungarian: 1887–1985).

Chapter 9
Existence and Definitions

In this chapter we take a closer look at existence statements. These statements assert that there exists a quantity x that has a certain property $\mathscr{P}$. In symbols:

$$\exists x \in X, \quad \mathscr{P}(x)$$

where X is a set and $\mathscr{P}$ is a predicate over X. For example, the statement

The integer 10^9 is the sum of two primes

asserts the existence of a pair of prime numbers with a given sum (what are X and $\mathscr{P}$ in this case?).

We shall discuss proofs of existence and the connection between existence and definitions.

9.1 Proofs of Existence

To prove that something exists it is not necessary that the object in question be constructed explicitly. For example, Euclid's proof of the infinitude of primes (Sect. 7.5) establishes the existence of infinitely many things without producing any of them. By contrast, we proved that Euler's polynomial $n^2 + n + 41$ is composite for infinitely many values of n by exhibiting such values explicitly (Eq. 7.8).

Accordingly, existence proofs are said to be **constructive** if an explicit construction is given, and **non-constructive** if no explicit construction is given. The constructive method may be described as a two-stage process:

(WHAT?) Identify an element x of X.
(WHY?)   Show that $\mathscr{P}(x)$ is TRUE.

The format of the proof should make clear which part is the WHAT and which part is the WHY.

We begin with a constructive existence proof [31, p. 13]:

Theorem. *There is a real number α such that $\alpha^2 = 2$.*

PROOF. Draw a square of side 1.

(WHAT?) Let α be the length of a diagonal of the square.
(WHY?) Then by Pythagoras, $\alpha^2 = 2$. □

This minimalist proof takes for granted that the length of a diagonal of the unit square is a real number. A justification of this step requires the tools of analysis.

Let us re-visit the counterexample (7.8) concerning Euler's polynomial.

Theorem. *There are infinitely many integers n for which $p(n) = n^2 + n + 41$ is composite.*

PROOF.

(WHAT?) Let $n = 41k$, for $k \in \mathbb{N}$.
(WHY?) Then $p(41k) = 41(41k^2 + k + 1)$, not a prime. □

In this proof the WHAT part identifies an infinite sequence of integers.

Theorem. *If two integers are the sum of two squares, then so is their product.*

This statement is an implication; the existence statement in the deduction is conditional to another existence statement in the hypothesis. We give a concise constructive proof, where an eloquent formula supplies at once the WHATs and the WHYs.

PROOF. This statement follows from the algebraic identity

$$(j^2 + k^2)(m^2 + n^2) = (jm + kn)^2 + (km - jn)^2.$$ □

Many non-constructive existence proofs employ contradiction. One assumes that the object being defined does not exist, and derives a false statement. It is not surprising that an argument of this kind could fail to provide information about the object in question.

However, not all non-constructive existence proofs use contradiction.

Theorem. *There is an irrational number a such that $a^{\sqrt{2}}$ is rational.*

PROOF. Consider $\sqrt{2}^{\sqrt{2}}$. We have two cases:
CASE I: $\sqrt{2}^{\sqrt{2}}$ is rational. Then $a = \sqrt{2}$ is as required.
CASE II: $\sqrt{2}^{\sqrt{2}}$ is irrational. Then

$$\left(\sqrt{2}^{\sqrt{2}}\right)^{\sqrt{2}} = (\sqrt{2})^{\sqrt{2}\sqrt{2}} = (\sqrt{2})^2 = 2,$$

so $a = \sqrt{2}^{\sqrt{2}}$ is as required. □

9.1 Proofs of Existence

This is a proof by cases—see Sect. 7.2. We note that we aren't given a WHAT. There are two different WHATs, and we have no idea which of them works for the WHY.

Our next example is a well-known existence statement, **Dirichlet's box principle** (or the **pigeon-hole principle**).

Theorem. *Given n boxes and m objects in them, if $m > n$, then at least one box contains more than one object.*

The proof of this rather self-evident implication is necessarily non-constructive. We use contradiction, namely the **both ends method** described in Sect. 7.5. It's easy to predict where contradiction will lead us.

PROOF. Let m and n be integers, and let m_j be the number of objects in the jth box. Assume that $m > n$ but $m_j \leqslant 1$, for $j = 1, \ldots, n$. Then

$$m = \sum_{j=1}^{n} m_j \leqslant \sum_{j=1}^{n} 1 = n$$

giving $m \leqslant n$, contrary to the assumption. □

In this proof we are not given the WHAT, and hence there cannot be a WHY step either. The proof begins by introducing the symbol m_j; this simple but important step makes the rest immediate.

The non-constructive proof of the following statement employs Dirichlet's box principle.

Theorem. *In any set of integers with more than n elements there must be two integers whose difference is divisible by n.*

PROOF. Let n be given, and let S be a set of m integers, with $n < m$. We divide each element of S by n, obtaining m integer remainders. These remainders can assume at most n distinct values, and therefore two elements of S, say x_i and x_j, must have the same remainder, by Dirichlet's box principle. But then $x_i - x_j$ gives reminder zero when divided by n, as desired. □

Solving equations is a quintessential mathematical task. In some cases we may be looking for a particular number, which means that we are after a constructive existence proof. In other cases we may be satisfied by a proof that a solution exists at all, or that it exists in a specified range of values of the argument. Let us examine a constructive and a non-constructive existence proof of the same statement.

Proposition. *The equation $x^4 - 10x^2 + 1 = 0$ has four distinct real solutions.*

CONSTRUCTIVE PROOF. Let $f(x) = x^4 - 10x^2 + 1$.
(WHAT?) Let $x_{\pm} = \sqrt{3} \pm \sqrt{2}$; then $x_{\pm}$ are real numbers.
(WHY?) We compute, using the binomial theorem

$$f(x_\pm) = (\sqrt{3} \pm \sqrt{2})^4 - 10(\sqrt{3} \pm \sqrt{2})^2 + 1$$
$$= 9 \pm 12\sqrt{6} + 36 \pm 8\sqrt{6} + 4 - 30 \mp 20\sqrt{6} - 20 + 1$$
$$= 0$$

where the top and bottom signs in $\pm$ and $\mp$ match. The above calculation shows that $x_\pm$ are roots of f. Since the real numbers $x_\pm$ are positive and distinct, and f is an even function, we conclude that $-x_\pm$ are also roots of f. □

This constructive existence proof involved 'guessing' the solutions, and then verifying that the guess was correct.[1]

NON-CONSTRUCTIVE PROOF. Let $f(x) = x^4 - 10x^2 + 1$. The computation

$$f(0) = 1 \quad f(1) = -8 \quad f(4) = 97,$$

shows that f changes sign twice in the interval $(0, 4)$. Because f is continuous, there exist two distinct real numbers x_-, x_+ in the open intervals $(0, 1)$ and $(1, 4)$, respectively, at which the function f vanishes. The real numbers $-x_\pm$ are also roots of f, because f is even. The four numbers $\pm x_\pm$ are clearly distinct. □

The key argument in the proof was inferring the existence of a root from a change of sign of a continuous function. This is the **intermediate value theorem**, a non-constructive existence theorem of real analysis [26, Theorem 4.4]:

Theorem. *Any real-valued continuous function f on a closed interval $[a, b]$ assumes every value between $f(a)$ and $f(b)$.*

Our non-constructive proof provides some information about the solutions, in the form of bounds. These bounds can be sharpened by doing some extra work; e.g., with f as above we have

$$f\left(\frac{7}{22}\right) = \frac{-503}{234256} \quad f\left(\frac{6}{19}\right) = \frac{1657}{130321}$$

and hence

$$\frac{6}{19} < x_- < \frac{7}{22}.$$

Considering that $7/22 - 6/19 < 3 \times 10^{-3}$, our knowledge of the solution x_- has increased markedly. We see that the distinction between a constructive and a non-constructive proof is blurred.

A non-constructive existence statement that provides some information about an object—typically a number—in the form of bounds, is said to be **effective**.[2] The

---

[1] There was no guess here: first I chose x_+, and then I derived the equation of which x_+ is a solution.
[2] Some authors use 'effective' to mean 'constructive'.

above non-constructive proof is effective. Likewise, the estimate of the size of the nth prime number given by Theorem 8.1 (Sect. 8.4) may be characterised as an effective version of Euclid's theorem.

In the mathematics literature the meaning of 'constructive proof' is sometimes extended to include methods that, while not providing an actual object, provide an algorithm for constructing the object.

9.2 Unique Existence

Unique existence is a stronger version of existence:

There is exactly one element of X having property $\mathscr{P}$.

where, as usual, X is a set and $\mathscr{P}$ is a predicate over X. For example:

The equation $f(x) = 0$ has exactly one solution.
The sets A and B have only one point in common.
The function g has precisely one stationary point.
The matrix M has a unique real eigenvalue greater than 1.

The expressions 'exactly', 'only', 'precisely' differentiate unique existence from existence.

Unique existence is a hidden conjunction; the conjuncts are

EXISTENCE: *There is one element of X with property $\mathscr{P}$.*
UNIQUENESS: *If $x, y \in X$ have property $\mathscr{P}$, then $x = y$.*

The corresponding symbolic expression is overloaded:

$$(\exists x \in X, \ \mathscr{P}(x)) \land (\forall x, y \in X, \ (\mathscr{P}(x) \land \mathscr{P}(y)) \Rightarrow (x = y)) \tag{9.1}$$

and to simplify it we introduce a new quantifier $\exists!$ denoting unique existence:

$$\exists! x \in X, \ \mathscr{P}(x).$$

For example, given a function $f: X \to Y$, the concise expression

$$\forall y \in Y, \ \exists! x \in X, \ f(x) = y$$

states that f is bijective.

In Chap. 8 we proved (twice) that any integer n greater than 1 is a product of primes. For each n, this statement asserts the existence of an unspecified number of primes. The **fundamental theorem of arithmetic** upgrades existence to unique existence.

Theorem. *Every natural number greater than 1 may be written as a product of prime numbers. This representation is unique, apart from re-arrangements of the factors.*

The expression 'apart from re-arrangements of the factors' gives the impression of lack of uniqueness. This is not the case: the theorem states that there exists a unique **multiset** of prime numbers (see Sect. 2.1) with the property that the product of its elements is the given natural number. Alternatively, there is a unique **set** of pairwise co-prime prime-powers with the same property.

In a proof of unique existence the conjuncts are normally proved separately, in either order.

Theorem. *The identity element of a group is unique.*

The group axioms were given in Sect. 7.1. The existence of an identity element does not require a proof, being an axiom (axiom G3). We only have to prove uniqueness.

PROOF. Let us assume that a group has two identity elements, $\Diamond_1$ and $\Diamond_2$. Applying the axiom G3 to each identity element, we obtain

$$\Diamond_1 \odot \Diamond_2 = \Diamond_1 \qquad \Diamond_1 \odot \Diamond_2 = \Diamond_2$$

and hence $\Diamond_1 = \Diamond_2$. □

Theorem. *For every set X and subset Y, there is a unique set Z such that $Y \cup Z = X$ and $Y \cap Z = \emptyset$.*

PROOF. Let $Z = X \smallsetminus Y$. By construction, $Y \cap Z = \emptyset$. Since $Y \cup Z$ is a subset of X, and every element of X belongs to either Y or Z, we have $Y \cup Z = X$. So a set Z with the required properties exists.

To prove uniqueness, assume that Z is a set with the stated properties. Then, since $Y \cup Z = X$, Z must be a subset of X, and it must contain all elements of X that aren't in Y. So Z must contain $X \smallsetminus Y$.

Since $Y \cap Z = \emptyset$, the set Z can't contain any elements of X that are in Y. So Z must be $X \smallsetminus Y$. □

The existence part of this proof is constructive. For this reason, the proof of uniqueness does not establish the second conjunct in (9.1) explicitly. Rather than considering two objects with the stated properties and showing that they are the same, we consider a single object and show that it is the same as the one defined in the existence part.

Theorem (Fermat). *The only natural numbers x for which $1 + x + x^2 + x^3$ is a square are $x = 1$ and $x = 7$.*

Proving existence is immediate: $1 + 1 + 1^2 + 1^3 = 2^2$, $1 + 7 + 7^2 + 7^3 = 20^2$. Uniqueness is much harder, see [12, pp. 38, 382].

9.3 Definitions

Definitions are closely related to unique existence. When we define a symbol or a name, we must ensure that this quantity actually exists, and that it has the stated properties. A definition may identify a unique quantity (*Let a be the length of a diagonal of a regular pentagon with unit area*), or specify membership to a unique non-empty set (*Let p be a prime such that p + 2 is also prime*). If these conditions are met, then our object is **well-defined**; otherwise it is **ill-defined**.

Let us begin with definitions of sets. Here there is a distinctive safety net: if the conditions imposed on a set are too restrictive, then this set will be empty rather than ill-defined. Nonetheless, we should be alert to this possibility, because the consequences of an empty definition could be more serious than formally correct nonsense. For example, consider the following definition:

> Let $\mathcal{M}$ be the set of 2 by 2 integral matrices with odd entries and unit determinant.

The set $\mathcal{M}$ is empty because the determinant of a matrix with odd entries is an even integer, and so it cannot be 1. It's easy to run into trouble now: *Let $M \in \mathcal{M}$*.

Sets defined by several conditions are commonplace; for instance, the set of solutions of a system of simultaneous equations is the intersection of the solution sets of the individual equations. A considerate definition should flag the possibility of an empty intersection:

> Let $A_1, \ldots, A_n$ be sets, and let A denote their (possibly empty) common intersection

$$A = A_1 \cap A_2 \cap \cdots \cap A_n. \tag{9.2}$$

For $n > 2$, the set A is well-defined, because the intersection operator is associative. For $n = 1$ the Formula (9.2) is, strictly speaking, unspecified, but the interpretation $A = A_1$ is quite natural, so a separate treatment is unnecessary. One could even stretch this definition to mean $A = \emptyset$ if $n = 0$, although a clarifying remark would be needed in this case.

Next we consider an innocent-looking integer sequence whose existence is more delicate than expected.

> Let q_n be the smallest prime number with n decimal digits.

Does such a prime exist for all n? Let's look at the first few terms of the sequence:

$$(q_n)_1^\infty = (2, 11, 101, 1009, 10007, \ldots). \tag{9.3}$$

In all cases, the prime q_n lies just above 10^{n-1}, and it seems unavoidable that there is at least one prime between 10^{n-1} and 10^n.

However, given that there are arbitrarily large gaps between consecutive primes,[3] an argument is needed to rule out the possibility that a large gap could include all integers with n digits, for some n. Such an argument is not elementary, and we shouldn't keep the reader pondering on this. So the definition of the sequence (q_n) should incorporate a remark or a footnote of the type:

This sequence is well-defined due to Bertrand's postulate[4]: for all $n \geqslant 1$ there is at least one prime p such that $n < p \leqslant 2n$ [15, p. 343].

A function definition $f: A \to B$ requires specifying two sets, the domain A and the co-domain B, as well as a rule that associates to every point $x \in A$ a unique point of $f(x) \in B$. For the function f to be well-defined, we must ensure that the specification of A and of the rule $x \mapsto f(x)$ do not contain any ambiguity.

By contrast, there is flexibility in the specification of the co-domain B, in the sense that any set containing $f(A)$ may serve as a co-domain. Formally, different choices of B correspond to different functions, although such distinctions are often unimportant. But then why don't we always choose $f(A)$ as co-domain? This would have the advantage of making every function surjective. The problem is that we may not know what $f(A)$ is, or the description of $f(A)$ may be exceedingly complicated.

For instance, let us return to the sequence (q_n) given in (9.3). Writing

$$q_n = 10^{n-1} + a_n \qquad (9.4)$$

we find

$$(a_n)_1^\infty = (1, 1, 1, 9, 7, \ldots).$$

Let us now define the function

$$f: \mathbb{N} \to \mathbb{N} \qquad f(n) = a_n.$$

This function is well-defined but we have limited knowledge of the image $f(\mathbb{N})$. It can be shown that $f(\mathbb{N})$ cannot intersect any of the sets $2\mathbb{N}$, $2 + 3\mathbb{N}$, and $5\mathbb{N}$, but we don't know if there are other constraints—see Exercise 10.4.3.

9.4 Recursive Definitions

A recursive definition is a form of definition connected to the principle of induction. With this method it is possible to define very complex objects—sequences, typically—from minimal ingredients.

---

[3] For $n > 1$, the integer $1 + n!$ is followed by $n - 1$ composite integers. (Why?)
[4] Conjectured by Joseph Bertrand (French: 1822–1900) in 1845. Proved by Pafnuty Chebyshev (Russian: 1821–1894) in 1850.

9.4 Recursive Definitions

We have pointed out that a sequence (a_k) corresponds to a function over $\mathbb{N}$, whereby a_k represents the value of the function at the point $k \in \mathbb{N}$. The existence of such a function does not necessarily imply that its values are easily recoverable from the definition of the sequence. For example, we don't know any 'useful' way to express the kth prime number as an explicit function of k. In absence of an explicit formula for the elements of a sequence, we seek to represent a_{k+1} in terms of a_k. These are **recursive sequences**, which are defined by two data: the first term of the sequence, called the **initial condition**, and the rule which determines a term from the previous one.

For instance, the **factorial sequence** $n! = 1 \cdot 2 \cdot 3 \cdots n$, can be defined recursively as follows:

$$0! = 1 \qquad (k+1)! = (k+1) \cdot k! \qquad k \geq 0. \tag{9.5}$$

This definition, in effect, specifies the order in which the product is computed. Likewise, the nth derivative $g^{(n)}$ of a function g is defined by the recursive rule

$$g^{(0)}(x) = g(x) \qquad g^{(k+1)}(x) = \frac{d}{dx} g^{(k)}(x) \qquad k \geq 0.$$

More generally, given any set X and any function $f: X \to X$, we define a recursive sequence (a_k) of elements of X as follows:

$$a_1 \in X \text{ given}; \qquad a_{k+1} = f(a_k), \quad k \geq 1. \tag{9.6}$$

The first term of the sequence (the initial condition) is given. Because f is a function, if a_k is well-defined for some $k \geq 1$, then so is a_{k+1}. The induction principle then ensures that the whole sequence is well-defined. Since distinct initial conditions in X give rise to distinct sequences, there are as many sequences as there are elements of X: plenty of sequences from just one function!

The term a_{k+1} of a recursive sequence may depend on $d \geq 1$ preceding terms, not just the last one:

$$a_{k+1} = f(a_k, a_{k-1}, \ldots, a_{k-d+1}). \tag{9.7}$$

In this case we have $f : X^d \to X$, and d is called the **order** of the sequence. In a recursive sequence of order d, the first d terms of the sequence must be supplied explicitly because they don't have the required number of preceding terms. These sequences have d initial conditions and they are well-defined by virtue of strong induction.

The **Fibonacci sequence**[5] is a second-order sequence:

$$a_0 = 1; \qquad a_1 = 1; \qquad a_{k+1} = f(a_k, a_{k-1}) = a_k + a_{k-1}, \quad k \geq 1.$$

---

[5] Leonardo Pisano, known as Fibonacci (Italian: 1170–ca.1240).

We find

$$(1, 1, 2, 3, 5, 8, 13, 21, 34, 55, 89, 144, \ldots).$$

Because the term $a_{k-1} = a_{k+1} - a_k$ is a well-defined function of the following two terms, we can extend the Fibonacci sequence backwards to obtain a doubly-infinite sequence:

$$(\ldots, -21, 13, -8, 5, -3, 2, -1, 1, 0, 1, 1, 2, 3, 5, 8, 13, 21, \ldots).$$

A recursive sequence that can be extended backwards is said to be **invertible**. Thus the Fibonacci sequence is invertible.

Recursive sequences may be quite unpredictable, as the following example illustrates. Let

$$f : \mathbb{N} \to \mathbb{N} \qquad x \mapsto \begin{cases} x/2 & \text{if } x \text{ is even} \\ 3x + 1 & \text{if } x \text{ is odd.} \end{cases}$$

We consider the first-order recursive sequence associated with this function:

$$x_0 = n \qquad x_{t+1} = f(x_t), \quad t \geq 0. \tag{9.8}$$

The initial condition $x_0 = 1$ leads to a **periodic sequence**

$$(1, 4, 2, 1, 4, 2, 1, \ldots) \tag{9.9}$$

consisting of indefinite repetitions of the pattern $(1, 4, 2)$. If instead we start with $x_0 = 7$, we find

$$(7, 22, 11, 34, 17, 52, 26, 13, 40, 20, 10, 5, 16, 8, 4, 2, 1, 4, 2, 1, \ldots)$$

so, after an irregular initial excursion, the sequence settles down to the same periodic pattern as the previous sequence. The analysis of the sequences (9.8) for general initial condition n presents formidable difficulties, which are distilled into a famous unsolved problem, the so-called '$3x+1$' **conjecture** [43]:

Conjecture. *For any n, the sequence (9.8) contains 1.*

This conjecture states that all these sequences are **eventually periodic** and that they reach the same periodic pattern.

We note that any open conjecture leads to objects whose existence is at present undecidable, see Exercises 9.4 and 10.4.

9.5 Wrong Definitions

As we did in Sect. 7.7 for logical arguments, we consider here some common mistakes made in definitions. A faulty definition may imply that the object being defined does not exist at all, or that it is not the one we had in mind, or that there is more than one object that fits the description.

We begin with an incorrect function definition which can be put right in several ways.

WRONG DEFINITION. *Let f be given by*:

$$f: \mathbb{Q} \to \mathbb{Q} \qquad f(x) = x^2 y + 1.$$

The mistake is plain but fatal: the expression $f(x)$ involves the unspecified quantity y. As defined, f is not a function. There are many legitimate interpretations of what this formula could mean, each resulting in a different function.

1. We re-define the domain, supplying the missing information via a second argument.

$$f: \mathbb{Q}^2 \to \mathbb{Q} \qquad f(x, y) = x^2 y + 1.$$

In this definition y is also rational; clearly there are other possibilities.

2. The missing argument is regarded as a **parameter** (see Sect. 6.1).

$$f_\lambda: \mathbb{Q} \to \mathbb{Q} \qquad f_\lambda(x) = \lambda x^2 + 1 \qquad \lambda \in \mathbb{Q}.$$

We have highlighted the change in status of the variable y by turning it into a subscript and switching to the Greek alphabet. For every value of λ, we have a different function.

3. The missing argument is regarded as an **indeterminate**. In this case $f(x)$ is a polynomial in y, and we must re-define the co-domain of f accordingly.

$$f: \mathbb{Q} \to \mathbb{Q}[y] \qquad f(r) = r^2 y + 1.$$

The symbol $\mathbb{Q}[y]$ denotes the set of all polynomials in the indeterminate y with rational coefficients (see Sect. 3.1). To avoid confusion, we have replaced x with a symbol that normally is not used for an indeterminate, and which reminds us that this variable is rational.

In the second example, the existence of a function requires more conditions than those given. The definition relies on a hidden assumption, so that using it adds extra information (we say that the definition is *creative*).

WRONG DEFINITION. *Let A and B be sets, and let $f, g : A \to B$ be functions. We define the function $h = f + g$ as follows*:

$$h: A \to B \qquad x \mapsto f(x) + g(x).$$

The hidden assumption is that the elements of the co-domain can be added together. However, B may be a set where addition is not defined (e.g., f and g are predicates) or which is not closed under addition (e.g., B is an interval). We give a more restrictive definition without hidden assumptions.

DEFINITION. *Let A be a set, let B be a set closed under addition, and let f, g: $A \to B$ be functions.* ...

In the next example we attempt to extend to modular arithmetic the concept of the reciprocal of an element, but our definition has a hidden assumption.

WRONG DEFINITION. *Let m be a natural number, and let a be an integer. If $[a]_m \neq [0]_m$, then we define the multiplicative inverse $[a]_m^{-1}$ as follows:*

$$[a]_m^{-1} \stackrel{\text{def}}{=} [b]_m \quad \text{where} \quad [a]_m[b]_m = [1]_m.$$

Take $m = 6$ and $a = 3$. We have $[3]_6 \neq [0]_6$, but the equation $[3]_6[x]_6 = [1]_6$ has no solution. What went wrong? We have assumed that the implication

$$\text{if} \quad a \neq 0 \quad \text{and} \quad ab = ac, \quad \text{then} \quad b = c,$$

which is valid for real or complex numbers, is also valid for congruence classes. Requiring that $[a]_m \neq [0]_m$, namely that a is not divisible by the modulus is not enough; for this implication to hold, we must assume that a is co-prime to the modulus.

DEFINITION. *Let m be a natural number and let a be an integer co-prime to m. We define the multiplicative inverse $[a]_m^{-1}$ of $[a]_m$ as follows...*

In the following definition both existence and uniqueness are problematic.

WRONG DEFINITION. *Let $p \in \mathbb{Z}[x]$, and let z_p be the root of p having smallest modulus.*

If p has degree zero, then z_p does not exist, so this definition relies on the hidden assumption that p has positive degree. If z_p exists, then it may still not be unique (e.g., $p = x^2 + 1$), but the use of the definite article ('the root') implies that it is. Requiring that p be non-constant solves the existence problem. The way we address uniqueness depends on the context. If we merely require that no root of p has smaller modulus than z_p, then we don't need uniqueness.

DEFINITION. *Let $p \in \mathbb{Z}[x]$ with $\deg(p) > 0$, and let z_p be a root of p with smallest modulus.*

This definition does not identify z_p; it tells us that z_p is a member of a well-defined non-empty set, but this set may have more than one element. If, on the other hand, we require a specific complex number, then we must supply additional information.

9.5 Wrong Definitions

In the following definition we impose a condition on the argument of z_p, which removes any ambiguity.

DEFINITION. *Let p be a non-constant polynomial with integer coefficients, and let z_p be the root of p with smallest modulus. If there is more than one root with this property, we let z_p be the root with smallest non-negative argument.*

In our final example, the definition hides a subtle lack of uniqueness.

WRONG DEFINITION. *Let $x \in [0, 1]$, and let f be the function that associates to x its binary digits sequence*

$$f: [0, 1] \to \{0, 1\}^{\mathbb{N}} \qquad x \mapsto (c_1, c_2, \ldots),$$

where

$$x = \sum_{k=1}^{\infty} \frac{c_k}{2^k} \qquad c_k \in \{0, 1\}. \tag{9.10}$$

(The co-domain of f is the set of all infinite binary sequences, see Sect. 3.5.1.) What's wrong with this definition? Consider the rational numbers

$$x_n = \frac{1}{2^n} = \sum_{k=1}^{\infty} \frac{1}{2^{k+n}} \qquad n \geqslant 1.$$

If we write x_n in the form (9.10), then the above identity shows that there are two binary representations, namely

$$(\underbrace{0, \ldots, 0}_{n-1}, 1, 0, 0, 0, \ldots) \quad \text{and} \quad (\underbrace{0, \ldots, 0}_{n-1}, 0, 1, 1, 1, \ldots). \tag{9.11}$$

So the function f is not uniquely defined at these rationals and, more generally, at all rationals of the form

$$\sum_{k=1}^{n-1} \frac{c_k}{2^k} + x_n$$

where $c_1, \ldots, c_{n-1}$ are arbitrary binary digits.

We resolve this ambiguity by choosing consistently the first of the two digit sequences in (9.11).

DEFINITION. *Let $x \in [0, 1)$ be given by*

$$x = \sum_{k=1}^{\infty} \frac{c_k}{2^k} \qquad c_k \in \{0, 1\}.$$

Without loss of generality, we assume that all sequences (c_k) contain infinitely many zeros. Let now f be the function that associates to x its binary digit sequence

$$f\colon [0, 1) \to \{0, 1\}^{\mathbb{N}} \qquad x \mapsto (c_1, c_2, \ldots).$$

Exercise 9.1 Use Dirichlet's box principle to prove the following existence statements.

1. Given any five points in the unit square, there are two points whose distance apart is at most $1/\sqrt{2}$.
2. In any finite group of people, there are two people with the same number of friends.
3. Every recursive sequence of elements of a finite set is eventually periodic.

Exercise 9.2 Each function definition contains an error. Explain what it is and how it should be corrected.

1. $f\colon \mathbb{R} \to \mathbb{R} \qquad x \mapsto \dfrac{1}{x^2 + x - 1}$

2. $f\colon \mathbb{R} \to \mathbb{R} \qquad x \mapsto \sqrt{x^2 - 1}$

3. $f\colon \mathbb{Q} \to \mathbb{Z} \qquad \dfrac{a}{b} \mapsto |a - b|$

4. $f\colon \mathbb{Z} \times \mathbb{R} \to \mathbb{Z} \qquad (x, n) \mapsto \sum_{k=1}^{n} \lfloor kx \rfloor$

5. $f\colon \mathbb{Z}[x] \times \mathbb{N} \to \mathbb{N} \qquad (p, n) \mapsto \gcd(p, x^n)$

6. $f\colon \mathbb{Z}^2 \to \mathbb{Z} \qquad (m, n) \mapsto n\mathbb{Z} \cup m\mathbb{Z}$.

Exercise 9.3 The following definitions have several flaws. Explain what they are, hence write a correct, clearer definition.

1. Let $a = (b_1, b_2, \ldots)$ be a given sequence, with $b_k \in B$. We define the function

$$f\colon \mathbb{Z} \to B \qquad f(m) = \sum_{k=1}^{m} b_k^{-1}.$$

2. Let X be a subset of $\mathbb{Q}$, and let $f(X)$ be the number of integers in X, namely

$$f\colon \mathbb{Q} \to \mathbb{N} \qquad f(X) = \#(x \in X \cap x \in \mathbb{Z}).$$

3. Let l be a line in the cartesian plane, let $P, Q \in \mathbb{R}^2$ be the points of intersection of l with the coordinate axes, and let $F(l)\colon \mathbb{R}^2 \to \mathbb{R}$ be the function that gives the length of the segment joining P and Q.

Exercise 9.4 Exploit each conjecture to define a function whose existence cannot be decided at present.

1. The twin-primes conjecture (Sect. 7.6).
2. The Goldbach conjecture (Sect. 7.6).
3. The $3x + 1$ conjecture (Sect. 9.4).

Chapter 10
Writing a Thesis

The highlight of mathematical writing at university is the report on a final year project, called a **thesis** or **dissertation**. This document has a distinctive structure, in between a short book and a research paper. A thesis surveys a body of advanced literature and presents original work. At undergraduate level, usually the former outweighs the latter.

Writing a thesis is an irreplaceable experience in university education. This document can reveal a great deal about the author's knowledge, understanding, curiosity, and enthusiasm. Examination scripts do not convey so much information.

One doesn't learn how to write a thesis by reading an instruction manual. The thesis will take shape gradually, as a result of regular interactions between student and supervisor. The document's high-level organisation—number of chapters, outline of their content—is normally considered at a relatively early stage of the project. Planning will also suggest appropriate headings and will aid the sequencing of the arguments. Periodic reviews of work in progress may alter priorities, or even re-direct the research. The writing process culminates with the exercise of proof-reading, in itself a valuable, if painstaking, experience.

In this chapter I give an overview of the structure of a thesis, and then discuss selected aspects of writing—choosing the title, writing the abstract, compiling the bibliography—where analysing examples may be helpful. I also include a brief introduction to LaTeX, the software adopted by the mathematical publishing industry. Most theses are now typeset in LaTeX.

10.1 Theses and Other Publications

A thesis is a substantial document. Its length varies considerably, depending on level, topic, and institutional requirements. An undergraduate thesis could be some fifty pages long. There are also theses at master and doctoral level, called MSc and PhD theses respectively.[1]

---

[1] Acronyms for Master of Science and Philosophy Doctor.

A good undergraduate thesis may rise to the level of a master thesis. A doctoral dissertation is longer, has greater depth, and more original research. The brief guidelines given in this chapter apply to any thesis. For further reading, see Higham's book [19], which devotes two chapters to PhD theses and has an extended bibliography.

A thesis is subdivided into heading, body, and closing matters. The heading supplies essential information: title, abstract, table of contents, acknowledgements. One may also find lists of mathematical symbols, figures, and tables, which some institutions require explicitly. Title and abstract define the subject matter; we'll consider them in some detail in Sects. 10.2 and 10.3. LaTeX will take care of the menial task of compiling the table of contents. The acknowledgements is a paragraph where the author thanks people for their guidance and support (the supervisor, typically), for useful discussions, for pointing out things, or for allowing the author to use their text in the thesis.

The body contains the bulk of the material, organised into chapters, sections, and, if appropriate, sub-sections. The first chapter is the introduction, written so as to make the thesis self-contained. Here we find motivations, background material and key references, enough details to be able to grasp the main results, and a description of the content of the rest of the thesis.

The length of individual chapters may vary, but each chapter must be a substantial component of the whole, with a clear identity. A brief closing chapter is desirable if not compulsory. This is an opportunity to re-visit ideas and results from an informed perspective, discuss limitations of the work, identify open problems, and chart directions of future research.

Any accessory material which should not interfere with the main text (data tables, computer programs, tedious proofs, etc.) is confined to appendices. The last item in a thesis is the bibliography, listing the sources cited in the text; it'll be considered in Sect. 10.4.

There is a symbiotic relationship between a thesis and a research paper; the higher the level of the thesis, the stronger the bond. At doctoral level, research papers lead to a thesis, or are extracted from it after the thesis is completed. At undergraduate level this connection is necessarily more remote, yet a research element must be present in every respectable thesis.

But a thesis is not a research paper. A research paper communicates new results to specialists, invariably within tight space constraints. It is seldom self-contained: sketchy motivations, unexplained jargon, and terse citations are commonplace. A thesis and a paper also differ in format. A paper is shorter, divided into sections, not chapters, while its simpler navigation renders a table of contents unnecessary.

A thesis is not a book either. A book covers thoroughly a body of established material. It is longer than a thesis, and we expect it to be wholly self-contained. A book begins with a preface, which is neither an abstract nor an introduction, but a bit of both. It is rare to find a book without exercises in each chapter, and even rarer one without a subject index at the end. (Notably, the celebrated book [15] has neither.) A textbook, being conceived as teaching material, may offer solutions or hints to exercises. This is a modern trend: hardly any of the classic, timeless mathematics textbooks provided solutions or hints to exercises—a fact worth pondering on. Some advanced publications resemble theses: the *review articles*, published in dedicated research journals, and the *research monographs*, published as books, usually within a collection. These are a sort of advanced theses, which collect together previously published works. They usually appear when a new research area has reached maturity.

Most scholarly publications are specialised; they are not intended for the general public. So how narrow should the target audience of a thesis be? This is largely a question of personal taste. Conscious decisions must be made from the very beginning, because the title, the abstract, and the first paragraphs of the introduction will set the tone for the rest of the document.

Let us inspect the opening sentences of some theses and research papers, to see how strongly the character of the work is established with just few words. Later on, we shall analyse titles and abstracts from the same perspective.

Our first example is a master thesis (from [35], with minor editing).

> Let $GL(d, q)$ denote the general linear group of invertible $d \times d$-matrices over the finite field $\mathbb{F}_q$ with q elements. In computational group theory, it is of interest to calculate the order n of an element $A \in GL(d, q)$, i.e.,
>
> $$n = \min\{j \in \mathbb{N} : A^j = I\}$$
>
> where I is the $d \times d$ identity matrix. The integer d can be large, e.g., $d > 100$.

The style is formal, with definitions and symbols appearing straight away. The jargon (general linear group, finite field, order of a matrix, etc.) and the corresponding notation ($GL(d, q)$, $\mathbb{F}_q$, etc.) belong to standard introductory algebra. The author still provides brief explanations, which convey an impression of helpful, considerate writing. This document promises to be accessible to any mathematically mature reader, a desirable feature of any undergraduate thesis.

Our next example, from a thesis in geometry [16], illustrates a completely different expository strategy.

> In 1984, Schechtman et al. [SGBC84] announced that the symmetry group of an aluminium-manganese alloy crystal, produced by rapid cooling, was that of the icosahedron. Such a symmetry is not possible for a periodic structure in three dimensions. This discovery brought down a long-held assumption in crystallography, that the only structures with some sense of long-range order were periodic.

The thesis begins with an account of the surprising outcome of a physics experiment. (We shall learn about literature citation in Sect. 10.4.) The author tells a story, skilfully building some drama. The near complete absence of mathematical terms—formal definitions and symbols are given much later—makes this excerpt very inviting. An engaging, accessible style is welcome in any publication, at any level.

A thesis can be highly specialised. One could use the established dictionary of a particular research area, without explanations or apologies, from the very beginning. This choice undoubtedly simplifies the writer's task, but it creates problems for the reader, thereby reducing the readership basis. This is legitimate, as long as the jargon is used wisely to sharpen the exposition, not gratuitously to impress the reader.

The following opening passage is taken from a PhD thesis in the area of *ergodic theory* [3].

> Ergodic optimisation is a branch of the ergodic theory of topological dynamics which is concerned with the study of T-invariant probability measures, and ergodic averages of real-valued functions f defined on the phase space of a dynamical systems $T : X \to X$, where X is a compact metric space.

This sentence explains the meaning of the expression 'ergodic optimisation', which features in the thesis' title. To achieve this within a limited space, the author relies on the reader's familiarity with half a dozen advanced concepts. Three important symbols are also defined in the same sentence. Undeniably, this thesis is for specialists.

In research papers, extremes of specialisation are accepted, or at least tolerated. Some editors insist—wisely, in my view—that the author provide clear motivations and a minimum of background, but this requirement is by no means universal. Papers that begin with a dry, unmotivated list of definitions are not uncommon. The following opening paragraph of standard definitions appears unchanged in three related publications [28–30].

> Let K be an algebraic extension of $\mathbb{Q}_p$, and let $\mathcal{O}$ be its integer ring with maximal ideal $\mathcal{M}$, and residue field k. If $\overline{K}$ is an algebraic closure of K, we denote by $\overline{\mathcal{O}}$ and $\overline{\mathcal{M}}$ the integral closure of $\mathcal{O}$ in K and the maximal ideal of $\overline{\mathcal{O}}$, respectively.

In the following extreme example, the author disposes of definitions altogether [7].

> 1. **Introduction.** The notation used is that of [2]; in particular …

One cannot proceed without the cited publication (by the same author). Only a committed reader will accept such a blunt treatment.

10.2 Title

The title provides an essential description of the content of a document. The titles of research papers compete for attention with other titles in journals, bibliographical lists, databases, etc. They help the reader decide whether to read on. A thesis has a captive readership—the examiners—so this problem does not arise immediately. Nonetheless, a good title can do a lot to give the thesis its unique identity, so it's worth spending some time thinking about it.

A title is a short phrase without symbols. To get an idea of the wide range of possibilities, we now examine titles of various types of publications, and assess the suitability of the title for a thesis, possibly after amendments. Alongside each title we display the bibliographical data—see Sect. 10.4.

Introductory real analysis [A.N. Kolmogorov and S.V. Fomin, Dover Publications, New York (1970).] The term 'introductory' immediately calls to mind a textbook, so this term is out of place in a thesis' title. The shorter title *Real analysis* would be much too general for a thesis, as it defines a vast area of mathematics.

Making transcendence transparent. (Subtitle: *An intuitive approach to classical transcendental number theory.*) [E. Burger and R. Tubbs, Springer-Verlag, New York (2004).] Subtitles can be quite effective. In this example, the title has an appealing alliteration, while the subtitle spells out the title's promise. However, the title's pedagogical slant and broad take are again the signature of a textbook, not a thesis.

Algebraic aspects of cryptography [N. Koblitz, Springer-Verlag, Berlin (1998).] The subject area covered by this textbook appears to be narrower than in the previous examples, whereas in fact it is still rather broad. If this title were used in a thesis, an expert would probably expect a literature survey. A thesis with an original research component would require a more specific title.

The arithmetic-geometric mean of Gauss. [T. Gilmore, MSc thesis, Queen Mary, University of London (2012).] As the reference to Gauss suggests, this topic is classical, and an expert will know that the arithmetic-geometric mean is connected to many deep mathematical theories. So this is a good title for a thesis, even if there is no indication of what might be the original contribution.

On the zeros of polynomials with restricted coefficients [P. Borwein and T. Erdelyi, *Illinois J. Math.* **41** (1997) 667–675.] This is a rare instance of a research problem with a concise and non-technical description, which makes a good title. The preposition 'on' lends an authoritative tone to it, promising some general result.

Toward a theory of Pollard's ρ-method [Eric Bach, *J. Information and Computation* **90** (1991) 139–155.] A very eloquent title. It suggests that the method being considered was not supported by a theory, and that some progress has been made. This would be an ideal title for a thesis. The use of the symbol ρ is justified, because the aforementioned method is invariably associated with it.

A complete determination of the complex quadratic fields of class number one [H.M. Stark, *Michigan Math. J.* **14** (1967) 1–27.] The forceful, unambiguous title of this well-known publication announces the solution of a long-standing problem. Such an assertive style suits well a research paper with a strong result; the trimmed version *Quadratic fields of class number one* could be the title of a thesis.

Diophantine integrability [R.G. Halburd, *J. Phys A: Math. Gen.* **38** (2005) 1–7.] This title is short and effective. Each term has an established meaning in a specific area of mathematics, and the unexpected juxtaposition of the two terms suggests interdisciplinarity. This title would be perfect for a thesis.

Outer billiards on kites [R.E. Schwartz, volume 171 of *Annals of Mathematics Studies*, Princeton University Press, Princeton and Oxford (2009).] The title of this research monograph cleverly exploits double meanings. Both terms 'Outer billiards' and 'kites' are mathematical, but not many readers will be familiar with them. The unlikely significance that this phrase would have in ordinary English is bound to raise interest among a general mathematical audience.

Euclidean algorithms are Gaussian [V. Baladi and B. Vallée, *Journal of Number Theory* **110** (2005) 331–386.] Summarising a result with a sentence makes an effective title for a research article, but perhaps less so for a thesis. The variant *Probabilistic aspects of Euclidean algorithms* is not so specific, and would work better for a thesis.

Metric number theory: the good and the bad. [R.E. Thorn, PhD thesis, Queen Mary, University of London (2004).] An excellent thesis title, enlivened by a colon. After announcing the research area (metric number theory), a clever reference to an epic Western provides a hint to the specific topic (badly approximable numbers).

To persist or not to persist? [J. Holfbauer and S.J. Schreiber, *Nonlinearity* **17** (2004) 1393–1406.] This paper is about 'uniformly persistent vector fields', and the author has transformed the subject matter into a Shakespearean question. Such a self-conscious title must be supported by strong content.

It is easy to determine whether a given integer is prime, [Andrew Granville *Bulletin of the American Math. Soc.* **42** (2005) 3–38.] Another title-sentence, with the added bonus of being provocative, because primality testing is a notoriously difficult problem. A fair deal of self-confidence is needed to choose a title of this kind. (This paper won the author the Chauvenet prize, awarded by the Mathematical Association of America to the author of an outstanding expository article on a mathematical topic.)

10.3 Abstract

An abstract is a short summary of a document. If well-written, it complements the title and sustains the reader's attention. The abstract of a research paper is placed immediately after the title, on the front page. In a thesis, where space constraints are not so severe, the abstract usually appears on a separate page.

It is difficult to compose a convincing abstract before the work has been completed, understood, and written out. So the abstract—the first thing one reads—is typically the last thing to be written. The main results must be in evidence, expressed with precision and clarity; at the same time, specialised jargon should be reduced to a minimum, so as not to alienate potential readers. A delicate balance must be achieved between these conflicting requirements.

Abstracts are short, rarely exceeding 200 words. The use of symbols is to be reduced to a minimum, the space is confined and every word counts. The exercise of writing MICRO-ESSAYs (see Sect. 6.6) is good training for writing abstracts, as both tasks require similar rigour and discipline.

We now examine in detail abstracts taken from mathematical literature. They are less than optimal, and we seek to improve them. (The original text has been edited, mainly to protect anonymity.) We don't intend to produce templates of well-written abstracts—this would be futile, as no two abstracts are the same. Rather, we want to sharpen our analytical skills, learn how to spot problems, and find ways of solving them. For this purpose, understanding the mathematics is not essential. Indeed, dealing with some unknown words and symbols has a certain advantage, as one must pay attention to the writing's internal structure. We have met a

10.3 Abstract

similar situation in Chap. 7, when we learnt how to organise the beginning of a proof.

Our first example, from a research paper, exemplifies a rather common problem: the use of unnecessary symbols.

> ABSTRACT. Let F be a rational map of degree $n \geqslant 2$ of the Riemann sphere $\overline{\mathbb{C}}$. We develop a theory of equilibrium states for the class of Hölder continuous functions f for which the pressure is larger than $\sup f$. We show that there exists a unique conformal measure (reference measure) and a unique equilibrium state, which is equivalent to the conformal measure with a positive continuous density.

The fault is plain: the symbols F, n, and $\overline{\mathbb{C}}$ are introduced, but not used again in the abstract, breaking the first golden rule of symbol usage—see Sect. 6.2. Unused symbols distract rather than help. Even if these symbols are used again in the main text, the abstract is not the place where the notation is to be established, and these definitions should be deferred. On this account, the symbol f in the second sentence is used appropriately. Embedded in the second sentence we find the abbreviation 'sup', for 'supremum'. This abbreviation (much like 'max', for maximum, or 'lim', for limit) is meant for formulae, and should not be mixed with words. So we replace it by the full expression.

Re-writing the first sentence without symbols and amending the second sentence do not require any understanding of the mathematics. We leave the third sentence as it is.

> ABSTRACT. We consider rational maps of the Riemann sphere, of degree greater than 1. We develop a theory of equilibrium states for the class of Hölder continuous functions f for which the pressure is larger than the supremum of f. We show that …

The symbol f in the second sentence is not strictly necessary, and can easily be disposed of.

> We develop a theory of equilibrium states for the class of Hölder continuous functions for which the pressure is larger than the supremum of the function.

The next abstract belongs to a thesis. The author presents an accessible problem using limited jargon.

> ABSTRACT. The logistic map is a well-studied map of the unit interval into itself. However, if we treat x as a discrete variable, as is done in any computer, then every orbit is eventually periodic. Thus the aperiodic behaviour that the continuous map displays for some value of the parameter r cannot be obtained from computer simulation. We investigated the differences and the similarities between the dynamics of a continuous map and its discrete approximation. We found that the limit cycles of a discrete map follow the unstable periodic orbits of the corresponding continuous map.

There are unknown words (logistic map, orbit, limit cycle), but also sufficiently many familiar terms (map, unit interval, eventually periodic, etc.) to help us discern the subject matter. The first sentence says that this work is about the 'logistic map', presented as a well-known object of investigation. The second and third sentences motivate the study. The author introduces the symbol x, but we are not told what it is. We guess that it must represent the argument of the logistic map, namely a point in the unit interval. Furthermore, this symbol is not used again in the abstract, so it is unnecessary. The symbol r suffers from similar problems. It's a parameter, but we are not told which quantity depends on it. We guess that this is the logistic map, which is referred to as the 'continuous map', presumably meaning that its argument assumes a continuum, rather than a discrete set, of values. The symbol r is not used again either.

With two important ideas not in sharp focus, the main message of the thesis—if we discretise the domain of the logistic map, then we obtain new phenomena worth studying—gets a bit lost.

The closing sentence, which summarises the main findings, is vague. What does 'we found' mean? Presumably this is not a proof, otherwise the author would have stated that clearly. What does 'follow' mean? Presumably it indicates some form of convergence. Even if in the abstract there isn't enough room for a complete statement of the results, we must find a way to render these vague statements acceptable. Finally, the author employs the past tense (we investigated, we found). In an abstract, the present tense is more common. Based on the above considerations, we rewrite the abstract as follows.

> ABSTRACT. We consider the logistic map, a well-studied map of the unit interval which depends on a parameter. If the domain of this map is discretised, as happens in any computer simulation, then, necessarily, all orbits become eventually periodic. Thus the aperiodic orbits observed for certain parameter values no longer exist. We investigate differences and similarities between the original map and its discrete approximations. We provide evidence that the limit cycles of the discrete map converge to the unstable orbits of the original map, in a sense to be made precise.

The last expression states that the thesis contains experimental data or heuristic arguments that support a clearly formulated notion of convergence. (We hope that this is the case in the present thesis!)

Next we consider the opening sentences of an abstract of a research paper. The problems are more subtle here: there are no superfluous symbols, and we can't rely on an understanding of the mathematics!

> ABSTRACT. In this work we investigate properties of minimal solutions of multi-dimensional discrete periodic variational problems. A well-known one-dimensional example of such a problem is the Frenkel-Korontova model. We select a family ...

The opening sentence begins with the expression 'in this work'; this is redundant and should simply be deleted. Expressions of this type serve a purpose only if the present

10.3 Abstract

work is being juxtaposed to other works: 'In 1964, Milnor proved an estimate…In this work, we prove…'

The object of the investigation is described by a long string of attributes: 'multi-dimensional discrete periodic variational problem'. Such a flat arrangement of words requires a pause or a highlight. At this stage however, it's not clear how this should be done, so we read on.

The second sentence begins with an expression of the type 'an example of this is that'. We have already criticised this format in Sect. 1.3. As written, the emphasis is placed on the word 'example', but the terms 'one-dimensional' or 'Frenkel-Korontova model' surely are more significant. The author's intentions are now clearer: the term 'multi-dimensional' in the first sentence is contrasted with 'one-dimensional' in the second sentence, and a known example is given for clarification. To get across the message that this work generalises 'one-dimensional' to 'multi-dimensional', we shall isolate the latter in the first sentence and emphasise the former in the second.

> ABSTRACT. We investigate properties of minimal solutions of discrete periodic variational problems, in the multi-dimensional case. These generalise one-dimensional problems, such as the well-known Frenkel-Korontova model. We select a family …

In our final abstract some symbols are necessary.

> ABSTRACT. Let $f : X \to X$, $X = [0, 1)$, be an interval exchange transformation (IET) ergodic with respect to the Lebesgue measure on X. Let $f_t : X_t \to X_t$ be the IET obtained by inducing f to $X_t = [0, t)$, $0 < t < 1$. We show that
> $$X_{wm} = \{0 < t < 1 : f_t \text{ is weakly mixing}\}$$
> is a residual subset of X of full Lebesgue measure. The result is proved by establishing a Diophantine sufficient condition on t for f_t to be weakly mixing.

The overall content of the abstract is unclear, although we recognise several words introduced in Chaps. 2 and 4. These fragments of understanding are sufficient to clarify the logic of the argument.

The set X is the unit interval, and we have a function f of this set into itself. This function belongs to a certain class of functions, identified by the acronym IET. The author assumes that f has certain properties, specified by some jargon. Then a parameter t is introduced, and the same construction is repeated for a one-parameter family of functions f_t over sub-intervals X_t. The functions f_t, constructed from f via a process we need not worry about, are also of type IET. (Evidently the author believes that this assertion does not require justification.)

Using a Zermelo definition, the author introduces a subset X_{wm} of the open unit interval, determined by a property of f_t, called 'weak mixing'. The subscript wm in the symbol X_{wm} must be an abbreviation for this expression. The paper establishes some properties of the set X_{wm}. An indication of the argument used in the proof is then given.

Our first task is to remove some symbols, including the displayed equation. This isn't too difficult; for example, the symbol X_{wm} is never used. The first two sentences are formal definitions, which need to be made more readable. The acronym IET is useful, since the long expression it represents is used twice. However, the primary object is 'interval exchange transformation', not IET, so the latter should follow the former, not precede it. Finally, in the expression 'Diophantine sufficient condition' the two adjectives compete with each other, so we separate them out.

> ABSTRACT. Let f be an interval exchange transformation (IET) of the unit interval, ergodic with respect to the Lebesgue measure, and let f_t be the IET obtained by inducing f on the sub-interval $[0, t)$, with $0 < t < 1$. We show that the set of values of t for which f_t is weakly mixing is a residual subset of full Lebesgue measure. The result is proved by establishing a Diophantine condition on t, which is sufficient for weak mixing.

More examples of abstracts in need of improvement are given as exercises.

10.4 Citations and Bibliography

Scholarly writing invariably requires reference to other sources. This is carried out in two stages. First, the source is mentioned in the text in an abbreviated manner. Second, a full ordered list of all cited sources (books, journal papers, web pages, etc.) is presented at the end of the document, in a section headed 'Bibliography' or 'References'.

There are some conventions for structuring citations and for writing the bibliography. We begin with a passage involving references to books.

> We briefly review some properties of p-adic numbers, introduced by Hensel in 1908 [18]. Background material may be found in Gouvea's (2000) [14]; the arithmetical side of the theory is developed in Hasse (1980) [17, Chap. 7].

Each item enclosed in square brackets is a citation, which refers to a bibliographical entry. Citations should not intrude upon the text, and a sentence in which all citations have been removed should still read well. The citations [14] and [18] refer to books as a whole, whereas [17] refers to a specific chapter in a book. The targeted item within a book may also be a theorem [15, theorem 191], a page [15, p. 156], a range of pages ([15, pp. 156–158], [15, 156*ff*][2]), etc. The bibliographical data corresponding to the text above are organised as follows:

[14] F.Q. Gouvêa, *p-adic Numbers*, Springer-Verlag, Berlin (2000).

⋮

[17] H. Hasse, *Number Theory*, Springer-Verlag, Berlin (1980).

---

[2] The suffix *ff* to a page number means 'and the following pages'.

[18] K. Hensel, *Theorie der Algebraischen Zhalen*, Teubner, Leipzig and Berlin (1908).

The entries are usually ordered alphabetically by the author's surname. Each item contains the following information: label, author(s), title, publisher, publisher's home town, year of publication. In our case the labels are numbers (there are other possibilities, see below), and the book's title is typeset in italic font. In [17] the publisher is 'Springer-Verlag', located in Berlin. As the publishing industry becomes global, the publisher's address is becoming increasingly irrelevant and publishers with multiple addresses are commonplace. The book identifier International Standard Book Number (ISBN), often given, is helpful for searches in libraries, catalogues, and databases.

There are other ways to cite books, as in the following shorter variant of our paragraph.

We briefly review some properties of p-adic numbers, introduced in [18]. Background material may be found in [14]: the arithmetical side of the theory is developed in [17, Chap. 7].

To simplify the citations we have removed the author's name and the year of publication from the text. However, reading has become more laborious because the reference number conveys no information about the source. The following labelling method offers a compromise, providing essential information in abbreviated form.

We briefly review some properties of p-adic numbers, introduced in [Hen08]. Background material may be found in [Gou80]; the arithmetical side of the theory is developed in [Has80, Chap. 7].

Citations to journal papers follow the same rules given for books, although the bibliographical information is slightly different. Here is one example:

[32] P. Morton and J.H. Silverman, Periodic points, multiplicities, and dynamical units, *J. Reine Angew. Math.*, **461** (1995) 81–122.

The data to be supplied comprises: label, author, title of paper, name of journal, volume number, year of publication, page numbers. The title is in Roman font, whereas the name of the journal appears in slanted font (a variant of italics), usually in a standard abbreviated form. Thus *J. Reine Angew. Math.* stands for *Journal für die reine und angewandte Mathematik*.[3] The volume number is in boldface. Page numbers are normally given in full, but the style 361–7 (rather than 361–367) is allowed. As for books, there is an identifier for journals, called International Standard Serial Number (ISSN).

Other sources for citations are articles in electronic journals [10], articles in conference proceedings [33], books in collections [4], reprints of books [9], theses [3], etc. There are great variations in the style of a bibliography. The best thing is to look around for examples of bibliographical entries of the various types by different authors and publishers.

---

[3] This is the oldest mathematics journal still in existence, founded in 1826.

For unpublished documents on the web there is as yet no agreed format. The main information is the Uniform Resource Locator (URL), which identifies a web page uniquely, as in the following example:

[27] Text processing using LaTeX, Dept. of Engineering, University of Cambridge, `http://www.eng.cam.ac.uk/help/tpl/textprocessing/` (January 2014).

This item is a collection of resources for electronic typesetting (see Sect. 10.5). The title is followed by the name of the institution providing the service and the URL, typeset in 'typewriter' font. There is no reference to authors—too many to be listed here. A date is provided. Web pages don't always carry dates (this one does), and even if they do, there is no way of deciding whether the displayed date corresponds to the last revision. The best one can do in this situation is to state when the web source was last accessed.

We present variants of bibliographical entries for web resources:

[8] P. Cvitanović, R. Artuso, R. Mainieri, G. Tanner, G. Vattay, *Chaos: Classical and Quantum*, `chaosbook.org` (Niels Bohr Institute, Copenhagen 2012).

[21] W. Jung, Core entropy and biaccessibility of quadratic polynomials, preprint (2014), `arXiv:1401.4792v1`, 46 pages.

Item [8] is a web-book, and the citation format adheres to the explicit instructions provided on the site. Instead of the URL there is a *domain name* registered in the Domain Name System (DNS), which provides a more intuitive identification of a web resource. Reference [8] has a large number of authors; to mention the authors in the text, an abbreviation is necessary: 'In Cvitanović et al. [8]...'.[4] The web-book [8] has been available for several years, and it is gradually evolving; this document structure would impossible within the confines of traditional publishing.

The item [21] is an unpublished paper. It is stored in a large, open-access archive of preprints of scientific papers, called `arXiv`, supported by Cornell University Library. Each preprint in this archive has a unique 8-digit identifier, plus a version number (`v1`, in this example) which reflects the fact that papers may be updated. Most preprints in the `arXiv` will eventually become available in a more permanent form. So, when proof-reading bibliographical entries, one should check that the papers in the `arXiv` haven't been published. Some editors accept only citations to permanent and accessible material.

A word of warning. When quoting from the web, be aware that there is a lot of rubbish there, and most web pages have no guarantee of correctness. Also keep in mind that anyone looking for basic information on a subject (e.g., *abelian groups*) can—without any assistance—type the relevant heading in a search engine and access the corresponding entry in Wikipedia. For this reason, citations to a main on-line encyclopaedia tend to be superfluous, and one should consider more substantial sources.

---

[4] Abbreviation for the Latin *et alii*, meaning 'and others'.

10.4.1 Avoiding Plagiarism

There is no copyright in mathematical ideas, the law of intellectual property being primarily commercial. But there is copyright in using other people's words, pictures, and software. The act of publishing material produced by others as if it were your own is a serious academic offence, called **plagiarism**.

To respect copyright when using other people's words one must name the author, cite the source, and make clear which are the quoted words (either by quotation marks or by display). For extensive use of someone else's text (more than a paragraph, say), one must obtain the explicit permission of the copyright holder, and then insert a suitable remark in the acknowledgements section: 'I thank XXX for permission to quote YYY'.

The instant availability of free material on the web has increased considerably the risk of plagiarism caused by carelessness and naïvety rather than dishonesty. When cutting and pasting from other people's text, it is essential to keep a record of the source. A text stored without a note of its source could easily be used as our own text at a later time. As articles, books, and theses are now submitted in electronic form, it is possible to detect this type of plagiarism using specialised software.

10.5 TEX and LATEX

The programming language TEX, developed by D. E. Knuth [23], and its user-friendly offspring LATEX, written by L. Lamport [25], are the standard software in mathematical and scientific publishing. This book was typeset in LATEX.

This typesetting system is intended for text containing mathematical formulae. It also supports the high-level features needed to manage a complex document: generating tables of contents, numbering of chapters and equations, cross referencing, handling bibliographical data, etc.

Let us inspect a very short LATEX source file[5]:

```
\documentclass[12pt]{article}
\begin{document}
In 1735, Leonhard Euler proved the following
remarkable formula:
\begin{equation}
    \sum_{n=1}^\infty\frac{1}{n^2}=1+\frac{1}{4}
      +\frac{1}{9}+\frac{1}{16}+\cdots=\frac{\upi^2}{6}.
\end{equation}
\end{document}
```

---

[5] The source file for the *Mathematical Writing* book contains over half a million characters.

The information is stored in plain text. We must type it in a file, and then run the LATEX program, which generates a new file suitable for display, typically in PDF format.[6] The end result is the following text, placed at the top of an empty page with the page number 1 at the bottom:

In 1735, Leonhard Euler proved the following remarkable formula:

$$\sum_{n=1}^{\infty} \frac{1}{n^2} = 1 + \frac{1}{4} + \frac{1}{9} + \frac{1}{16} + \cdots = \frac{\pi^2}{6}. \tag{1}$$

The contrast between the awkwardness of the LATEX source and the neatness of the output may be off-putting at first. On closer acquaintance, it will instead become clear why this logical and powerful language has become so widespread.

In the source file we note frequent occurrences of the distinctive backslash symbol '\', followed by a word. These are LATEX commands, which instruct the program to do something. We see that, after an initial preamble, a pair of commands delimit the beginning and the end of the document. Within the document, another pair delimits a displayed equation, and within the equation various commands generate mathematical symbols and structures. Certain things occur automatically, such as the centering and numbering of the displayed equation, and the output is insensitive to the layout of the source code.

Of course, the software does a lot more than typesetting. Thus LATEX may be expanded by incorporating auxiliary *libraries* of commands for specific tasks, such as BIBTEX, which manages bibliographical information. Similarly, the appearance of a document may be altered by means of *macros*. These are libraries transparent to the user, which reset the layout parameters (font type and size, line and paragraph spacing, style of bibliographical list, etc.). Publishers use this facility to convert standard LATEX source files into documents with the desired appearance.

Learning LATEX takes some time, but if you plan to write a thesis this is a worthwhile investment. There is no shortage of introductory material: the Cambridge University LATEX page is a good place to start [27].

Exercise 10.1 The following abstracts are poorly written. Identify the problems, hence write an improved version.

1. In this project the author examines some properties of continued fractions (CF). In the beginning, definitions, notations, and basic results and theorems are shown. Periodic continued fractions and best approximations are examined subsequently in depth. We examined a number of applications to mathematics and astronomy.
2. The most representative example of a 2-dimensional area-preserving twist map is the standard map, which is studied. Orbits for which the momentum p grows linearly (plus a periodic function) are shown to exist, classified and determined numerically. These orbits are the accelerator modes. The linear stability of these orbits is determined. The range in the parameter values for which they exist is also determined.

---

[6] This is an established format for document representation.

3. We consider equal parameter generalized quadrangles, $GQ(s, s)$. All GQs of order 2 and 3, $GQ(2, 2)$ and $GQ(3, 3)$, are known. It is conjectured all the GQ of order five are known as well. The known GQ of order 5 is the symplectic GQ, $W(5)$. $W(5)$ along with its dual are conjectured to be the only GQs of order 5. The construction of a symplectic GQ is given and then used to construct the known GQ of order 5. Information about GQ was gathered, including some basic combinatorics, affine GQ, and incidence matrices in an attempt to prove the above conjecture.
4. Under the boundary condition on the initial value $T_k(0)$ ($T_k(0) > 0$; $k = 1, 2, \ldots$) that $T_k(0) \to 0$ ($k \to \infty$), we integrate the semi-infinite system of nonlinear differential equations $\dot{T}_k = 2T_k(T_{k+1} - T_{k-1})$, ($k = 1, 2, \ldots$; $T_0 = 0$) to obtain their general solution. We further investigate the asymptotic time behaviours of this general solution as $t \to \pm\infty$.

Exercise 10.2 This is the longest opening sentence of a mathematical abstract I managed to find in the literature [39].

By Morava's point of view on the stable homotopy category, the quotient in some sense associated to the filtration related to the height of formal group laws is studied by the category of modules over the function ring of the deformation space of the Honda group law of height n with the lift of the action of the automorphism group on the closed fibre through the Adams-Novikov spectral sequence.

1. Can you break it into shorter sentences?
2. Can you find a longer opening sentence in any mathematical publication?

Exercise 10.3 Explore the statement (4.31), and write a one-page report on your findings.

Exercise 10.4 This exercise proposes some MICRO-PROJECTS in experimental mathematics. You are given a sequence to study; your task is to generate numerically a sufficient number of terms, analyse the data, and then write a succinct report—three/four pages, say.

In the report you must define the problem, give a concise account of your findings, and synthesise them in a precise mathematical conjecture. Invariably, some plots will be necessary. You should provide at least one bibliographical item, and include the numerical code in an appendix.

For an introduction to experimental mathematics, see [42]; the book [5] is a good source of computational problems suitable for projects.

1. EULER'S POLYNOMIAL. Let $p(n) = n^2 + n + 41$ be Euler's polynomial (see Sect. 7.6), and let a_n be the number of prime values assumed by p over the first n natural numbers, that is,

$$a_n = \#\{k \in \mathbb{N} : k \leqslant n, \ p(k) \text{ is prime}\} \qquad n \geqslant 1.$$

Refine conjecture 3 of Sect. 7.6 by identifying non-trivial bounds of the form

$$f(n) \leqslant a_n \leqslant g(n) \tag{10.1}$$

to hold for all sufficiently large n. The functions f and g must be chosen so as to minimise the growth rate of the difference $g - f$.

2. EXPONENTIAL SUMS. Given a real sequence (a_k) we define a sequence (z_n) of complex numbers via the sum

$$z_n = \sum_{k=1}^{n} e^{2\pi i a_k} \quad n \geqslant 1.$$

The number z_n may be interpreted as the end-point of a walk on the complex plane, consisting of n steps of unit length. Study (z_n) for one of the following sequences (a_k):

a. $a_k = k^2/50$
b. $a_k = k^2/(50 + 1/100)$
c. $a_k = k \log(k)$.

Formulate a conjecture concerning the behaviour of z_n or $|z_n|$ as n tends to infinity.

3. PRIMES AND POWERS OF TEN. Let a_n be as in (9.4).

a. Prove that $f(\mathbb{N}) \subset A$, where

$$A = \mathbb{N} \setminus (2\mathbb{N} \cup (2 + 3\mathbb{N}) \cup 5\mathbb{N}). \tag{10.2}$$

b. Provide experimental evidence that $f(\mathbb{N})$ is unbounded.
c. Provide experimental evidence that $f(\mathbb{N}) = A$.

Solution to Exercises

Exercise 1.1

1. The number a is positive.
2. The only even prime is 2.
   (The integer 2 is the only even prime.)
3. If $x > 0$, then $g(x) \neq 0$.
4. We change sign to both sides of the equation.
5. The polynomial $x^2 + 1$ has no real roots.
   (The equation $x^2 + 1 = 0$ has no real solutions.)
6. Multiplying both sides by a negative value of x, the inequality is reversed.
7. All solutions are odd.
   (The solution set consists of odd integers.)
8. If $\sin(\pi x) = 0$, then x is an integer.
   (The condition $\sin(\pi x) = 0$ implies that x is an integer.)
9. A matrix is invertible if its determinant is non-zero.
10. This infinite sequence has fewer negative terms.

Exercise 1.2

1. There are three special cases.
3. It does not tend to infinity.
5. Therefore c^{-1} is undefined.
7. We square both sides of the equation.
9. The equation $x^2 = y^2$ represents (describes) two orthogonal lines.
11. All matrices in this set are invertible.
13. Purely imaginary means that the real part is zero.
15. Since x is zero, I can't divide by x.
17. The function f is continuous.
19. I found fewer solutions than I expected.
21. We prove Euler's theorem.
23. The asymptotes of this hyperbola are orthogonal.
25. The solution depends on s.

27. Thus $x = a$. (We assume that a is positive.)
29. Always remember to check the sign.

Exercise 2.1
BAD: Is 39 a prime number? [*Specific and insignificant.*]
GOOD: Why is 1 not a prime number?
BAD: What is $1/2 + 1/2^2 + 1/2^3$?
GOOD: Is a fraction the same as a rational number?
BAD: What is the real part of $2 + 3i$?
GOOD: What is the real part of i^i?

Exercise 2.2
BAD: The set of natural numbers less than 10.
GOOD: The set of the proper subsets of a finite set.

Exercise 2.3

1. $\{1 - 2n : n \in \mathbb{N}\}$
   ($\{n \in \mathbb{Z} : n < 0, 2 \not\mid n\}$).
3. $\{(n+1)/n : n \in \mathbb{Z}\smallsetminus\{0\}\}$.
5. $\{z \in \mathbb{C} : |z| \geqslant 1\}$.
7. Let $F(x, y) = 0$ be the equation of a circle with the stated property. The function F must be such that $F(0, 0) = 0$.
9. No such a line goes through the origin, so we represent lines as $ax + by + 1 = 0$. The tangency condition results in a relation between a and b.

Exercise 2.4

1. The set of rational points in the open unit interval.
3. The set of rationals whose denominator is a power of 2.
5. The imaginary axis in the complex plane, excluding the origin.
7. The set of integer pairs whose first component divides the second.
9. The set of points in a euclidean space whose coordinates have zero sum.

Exercise 2.5

1. The inverse of a function.
3. The value of a function at the reciprocal of the argument.
5. The composition of a function with itself.
7. The restriction of a function to the integers.
9. The image of the irrational numbers under a function.

Exercise 2.6

1. The quotient of two fractions is a fraction. Its numerator (denominator) is the product of the numerator (denominator) of the first fraction and the denominator (numerator) of the second fraction.

Solution to Exercises

3. Compute the cube of the natural numbers: 1, 8, 27, etc. Stop when either (i) you obtain your integer (which is then a cube); or (ii) you obtain a larger integer, in which case your integer is not a cube.
   If your integer is 1, then it's a cube; otherwise, perform its prime factorization. If the exponent of each prime factor is divisible by 3, then your integer is a cube.
5. Transform the equation of the circle so that one side is zero, and the other side has positive quadratic coefficients. Then substitute the coordinates of the point into the equation. If you get a negative value, then the point is inside the circle; otherwise it's outside (or on the boundary).
7. Form the matrix that has your vectors as rows (or columns). Then compute the determinant of the matrix, and verify that it is non-zero.
9. Two things must happen: (i) there is a point at which the value of the two functions is the same; (ii) at this point, the product of the derivatives of the two functions is equal to -1.

Exercise 2.7

1. Compute the value of the function at each point of the domain, collecting together the resulting values. Eliminate all duplicates, so as to turn this list into a set. Your function is surjective precisely when the cardinality of this set is the same as that of the co-domain.
   Compute the value of the function at each point of the domain, collecting together the resulting values. Now verify that each element of the co-domain also belongs to your list of values. If this is true, then your function is surjective; otherwise it isn't.

Exercise 2.8

1. Let $X = A^2 \setminus B^2$ and $Y = (A \setminus B)^2$. If $A \subset B$, then, necessarily, $A^2 \subset B^2$, and hence $X = Y = \emptyset$. If A and B are disjoint, then so are A^2 and B^2, and hence $X = Y = A^2$. We will show that in all other cases Y is a proper subset of X.
   We begin with an example. Let $A = \{1, 2\}$ and $B = \{1\}$. The pair $(2, 1)$ belongs to $A^2 \setminus B^2 = \{(1, 2), (2, 1), (2, 2)\}$ but not to $(A \setminus B)^2 = \{(2, 2)\}$, as desired. Let us generalise this argument. If A is not a subset of B, then there is an element a in $A \setminus B$, and if A and B are not disjoint, then we can find $b \in A \cap B$. Now, the pair $z = (a, b)$ belongs to X (because $z \in A^2$ and $z \notin B^2$) but doesn't belong to Y (because the second component of z is not in $A \setminus B$). We have shown that Y is a proper subset of X, as desired.

Exercise 2.11 First fix some ambient set. What are the domain and co-domain of such a function?

Exercise 2.12 Consider *infinite* sequences of non-negative integers, where the kth term represents the power of the kth prime in the prime factorisation of an integer.

Exercise 2.13

1. Let Γ be such a set. We identify a segment in Γ via its mid-point z (a pair of real numbers), length r (a positive real number), and orientation θ (an angle between 0 and π). We see that
$$\Gamma \sim \Gamma^* \quad \text{where } \Gamma^* := \mathbb{R}^2 \times \mathbb{R}^+ \times [0, \pi).$$

   Consider now the subset U of Γ consisting of all segments of unit length. Using our representation, we have
$$U \sim \{(z, r, \theta) \in \Gamma^* : r = 1\} \sim \mathbb{R}^2 \times [0, \pi).$$

   In the last equivalence, we have removed the idle variable r, because its value is fixed.
2. Consider the representation of a segment.
3. Consider infinite integer sequences.
4. Consider sums of the type
$$\sum_{k=n}^{\infty} d_k 10^{-k}$$
   where n is an integer, and (d_k) is the sequence of decimal digits.
5. Consider the binary digits of fractions.

Exercise 3.1

1. $a_1 + a_2 + a_3$
3. $a_2 + a_1 + a_0 + a_{-1}$
5. $a_0 + a_2 + a_3 + a_4 + a_6$.

Exercise 3.2

1. (i) An identity.
   (ii) An arithmetical identity, expressing a cube as the sum of three cubes.
3. (i) A chain of inequalities.
   (ii) Upper and lower rational bounds for the square root of 2.
5. (i) An inequality.
   (ii) An algebraic inequality in two unknowns.
   (The inequality defining the first and third open quadrants in the cartesian plane.)
7. (i) An equation.
   (ii) The cartesian equation of a parabola passing through the origin.
9. (i) An identity.
   (ii) The trigonometric formula for the sine of the difference of two angles.

Solution to Exercises

Exercise 3.3

1. (i) A set.
   (ii) The set of even integers that are not divisible by 4.
3. (i) A set.
   (ii) The set of rational points on the plane, with non-integer components.
   (The set of pairs of rational numbers that are not integers.)
5. Same as Exercise 2.4.11.
7. (i) A sequence.
   (ii) An infinite sequence of sets, each containing one and the same element.
9. (i) A finite sequence.
   (ii) A finite sequence, obtained by raising consecutive natural numbers to the same even power.

Exercise 3.4

1. (i) A function.
   (ii) The real function that adds 1 to its argument.
   (The function that performs the right unit translation on the real line.)
3. (i) An identity. (A functional identity.)
   (ii) The formula for the derivative of the product of two functions.
5. (i) An integral.
   (ii) The indefinite integral of a function of two variables, performed with respect to the first variable.
7. (i) An identity. (A definition.)
   (ii) The power series of the cosine.
9. (i) A finite product of functions. (An analytic expression.)
   (ii) The product of all the partial derivatives of a function of several variables.

Exercise 3.5

1. (i) A set.
   (ii) The intersection of the inverse images of the elements of a sequence of sets.

Exercise 4.1

1. $\exists x \in X, \ x \notin Y$
3. $\exists x, y \in X, \ x \neq y$

Exercise 4.2

1. $\exists k \in \mathbb{Z}, \ n = k^3$
3. $\exists x \in \mathbb{Q}, \ f(x) = 0$
5. $\forall x \in \mathbb{Z}, \ p(x) \neq 0$
7. $\exists y \in B, \ \forall x \in A, \ f(x) \neq y$
9. $\exists x, y \in A, \ f(x) \neq f(y)$

Exercise 4.3 In part (i) there is freedom in the choice of the ambient set: the smaller the set, the simpler the predicate. (But if the set becomes too small the implication disappears, see Exercise 4.10.3.)

1. (i) For all primes p, if $p > 2$, then p is odd. (TRUE)
   (ii) For all primes p, if p is even, then $p \leqslant 2$.
   (iii) For all primes p, if p is odd, then $p > 2$. (TRUE)
   (iv) There is a prime p such that $p > 2$ and p is even.
3. (i) For all positive fractions a/b, if a/b is reduced, then b/a is reduced. (TRUE)
   (For all fractions a/b, if a/b is positive and reduced, then b/a is reduced.)
   (ii) For all positive fractions a/b, if b/a is not reduced, then a/b is not reduced.
   (iii) For all positive fractions a/b, if b/a is reduced, then a/b is reduced. (TRUE)
   (iv) There is a positive fraction a/b such that a/b is reduced but b/a is not reduced.
5. See end of Sect. 4.5.
7. (i) For all odd integers n, if n is greater than 3, then one of $n, n+2, n+4$ is not prime. (TRUE)
   (For all integers n, if n is odd and greater than 3, then one of $n, n+2, n+4$ is not prime.)
   (ii) For all odd integers n, if $n, n+2$ and $n+4$ are all prime, then $n \leqslant 3$.
   (iii) For all odd integers n, if one of $n, n+2$ and $n+4$ is not prime, then $n > 3$. (FALSE)
   (iv) There is an odd integer n such that $n > 3$, and $n, n+2, n+4$ are all prime.
9. Make sure that hypothesis and conclusion are not contradictory.

Exercise 4.4 This is the definition of nephew:
x is a nephew of y := There is z such that x is a son of z and
$\qquad$ (z is a brother of y or z is a sister of y).

Thus to define nephew we must first define of brother and sister; in turn these will require other definitions.

Exercise 4.5

(a.1) The integer 5 has a proper divisor. (FALSE)
(a.3) Every natural number properly divides its square. (FALSE)
(b.1) The integer y has a proper divisor.
  (c) Consider part (b).

Exercise 4.6

1. The set of prime numbers, together with the integer 1.

Exercise 4.7 Use Theorem 4.3, Sect. 4.5.

1. (FALSE). $\exists n \in \mathbb{N}, \ 1/n \in \mathbb{N}$.
   The reciprocal of a natural number is not a natural number.
   There is a natural number whose reciprocal is also a natural number.
3. (TRUE). $\exists x, y \in \mathbb{R}, \ xy \neq yx$.
   The product of real numbers is commutative.
   The product of real numbers is not commutative.
5. (TRUE). $\exists m, n \in \mathbb{Z}, \ (m+n \in 1+2\mathbb{Z}) \wedge (m \notin 2\mathbb{Z} \wedge n \notin 2\mathbb{Z})$.
   If the sum of two integers is odd, then at least one summand is even.
   There are two odd integers whose sum is odd.

Exercise 4.9 Let $G = \{1, \ldots, n\}$. Think of the computation of the predicate as the evaluation of the entries of an $n \times n$ matrix, where the (i, j)-entry is the value of 'i loves j'. Then decide the order in which the entries are accessed, e.g., first by row, then by column.

1. Let $g = n$, say. If the sentence is true, then for all $k = 1, \ldots, n$ we must verify that $\mathcal{L}(k, g) = $ T (George is included), so in any case we need n function evaluations. If the sentence is false, then the minimum is 1 (we find $\mathcal{L}(1, g) = $ F at the first evaluation). The maximum is n (we find $\mathcal{L}(k, g) = $ T for all $k \neq n$, and $\mathcal{L}(g, g) = $ F).

Exercise 4.10

1. Use Theorem 4.1.
3. Consider the set of which $\mathscr{P}$ is the characteristic function.

Exercise 5.1 First translate symbols into words literally; then synthesise the meaning of the literal sentence. (It may be helpful to draw the graph of a function that has the stated property, and a function that hasn't.)

1. (i) The value of f at the origin is rational.
   (ii) $\exists r \in \mathbb{Q}, f(0) = r$.
3. (i) The function f is constant.
   (ii) $\exists y \in \mathbb{R}, \forall x \in \mathbb{R}, f(x) = y$.
5. (i) The function f vanishes at some integer.
   (ii) $\exists x \in \mathbb{Z}, f(x) = 0$.
7. (i) The function f assumes only rational values.
   (ii) $\forall y \in \mathbb{R}, f(x) \in \mathbb{Q}$.
9. (i) The integers and the natural numbers have the same image under f.
   (ii) $\forall y \in \mathbb{Z}, \exists x \in \mathbb{N}, f(x) = f(y)$.

Exercise 5.3

1. The zeros of f include all even integers.
3. The function f is identically zero for negative values of the argument.
5. The function is non-zero at all rationals, with the possible exception of 0.
7. The function f vanishes infinitely often on the natural numbers.

Exercise 5.4

1. $\forall x \in \mathbb{R}, f(x) \neq 0$.
3. $\forall x \in \mathbb{R}, f(x) = 0 \Rightarrow x = 0$.
5. $\forall x \in \mathbb{Z}, f(\pi x) = 0$.
7. $\forall \varepsilon \in \mathbb{R}^+, \exists x \in (-\varepsilon, \varepsilon) \setminus \{0\}, f(x) = 0$.

Exercise 5.5

1. (a) If $-f$ is increasing, then f is decreasing. (TRUE)
   (b) If $-f$ is not increasing, then f is not decreasing. (TRUE)
   (c) There is a decreasing function f such that $-f$ is not increasing.

3. (a) If f is monotonic, then $|f|$ is increasing.
   (FALSE, e.g., $f(x) = x$)
   (b) If f is not monotonic, then $|f|$ is not increasing.
   (FALSE, e.g., $f(x) = e^x$ for $x < 0$ and $-e^x$ for $x \geq 0$.)
   (c) There is a non-monotonic function whose absolute value is increasing.
5. (a) If f is surjective, then f is unbounded. (TRUE)
   (b) If f is not surjective, then f is bounded.
   (FALSE, e.g., $f(x) = \lfloor x \rfloor$.)
   (c) There is an unbounded function which is not surjective.

Exercise 5.6

1. (i) For all real functions f, if f is differentiable, then f is continuous. (TRUE)
   (ii) For all real functions f, if f is discontinuous, then f is not differentiable.
   (iii) Every continuous real function is differentiable. (FALSE)
   (iv) There is a real function f which is differentiable but not continuous.
3. (i) For all real functions f and g, if f and g are odd, then $f + g$ is odd. (TRUE)
   (ii) For all real functions f and g, if $f + g$ is not odd, then one of f or g is not odd.
   (iii) For all real functions f and g, if $f + g$ is odd, then f and g are odd. (FALSE)
   (iv) There are odd real functions f and g such that $f + g$ is not odd.

Exercise 5.8

1. A smooth bounded odd function, which vanishes infinitely often and approaches zero alternating sign as the argument tends to infinity.
3. This is a step function defined only for positive arguments. The steps have unit length, while their height increases monotonically.
5. A differentiable even function, with a global minimum at the origin and no maximum. For positive arguments the function increases monotonically, approaching a positive limit.

Exercise 5.10

1. The logarithm.
3. The identity.
5. The sine of the square of the argument.
7. The square of the arc-tangent.

Exercise 5.11

1. $\exists k \in \mathbb{N}, \ a_k \neq b_k$
3. $\forall n \in \mathbb{N}, \ \exists k \in \mathbb{N}, \ a_k \neq a_{k+n}$
5. $\forall k \in \mathbb{N}, \ \exists j \in \mathbb{N}, \ a_{k+j} < 0$
7. $\forall \varepsilon \in \mathbb{R}^+, \ \exists k \in \mathbb{N}, \ |a_k| < \varepsilon$
9. $\forall n \in \mathbb{N}, \ \exists k \in \mathbb{Z} \setminus \{0\}, \ a_{n+k} = a_k$.

Solution to Exercises

Exercise 6.2

1. When we say that the price of petrol, say, on Thursday, is 10% higher that the price on Wednesday, we mean that the Thursday price is equal to the Wednesday price plus one tenth of the Wednesday price, so it's $(1 + 1/10) = 11/10 = 1.1$ times the Wednesday price. Likewise, if we say that the Friday price is 10% lower than the Thursday price, that means it's the Thursday price minus one tenth of the Thursday price (*not* one tenth of the Wednesday price!); so the Friday price is $(1 - 1/10) = 9/10 = 0.9$ times the Thursday price. In all, the Friday price is the Wednesday price times 11/10 times 9/10, which is $99/100 = 0.99$ times the Wednesday price—nearly the Wednesday price but not quite.

2. Let $\mathbb{R}_{\geq 0}$ be the set of non-negative real numbers, representing the values of an observable (e.g., the price of a commodity). We consider the function

$$f_p : \mathbb{R}_{\geq 0} \to \mathbb{R}_{\geq 0}$$

where $f_p(x)$ is the result of varying x by a given percentage amount p, followed by a variation by $-p$. The parameter p is a real number, which we require to be in the range $-100 \leq p \leq 100$, to ensure that the process does not result in negative quantities, i.e., that $f_p(x) \in \mathbb{R}_{\geq 0}$.

A single change of x by $p\%$ corresponds to the mapping

$$x \mapsto x \left(1 + \frac{p}{100}\right).$$

If this transformation is followed by a percentage change with the opposite sign, we obtain

$$f_p(x) = x \left(1 + \frac{p}{100}\right)\left(1 - \frac{p}{100}\right) = x \left[1 - \left(\frac{p}{100}\right)^2\right].$$

For fixed p, this is a linear function of x, whose coefficient depends on $|p|$. If we exclude the trivial case $p = 0$, we have $f_p(x) < x$, which means that at the end of the process the value of x has become smaller. The decrease is maximal if $p = \pm 100$, in which case the final value is zero, irrespective of the initial value.

Exercise 6.5 Imagine that there are 100 boxes, instead of three; the presenter opens 98 boxes, all of them empty.

Exercise 6.7

1. You should identify at least three definitions.
2. Construct a sequence of intervals with given mid-point and length converging to zero.
3. The domain of a 'random variable' is the space of elementary outcomes. For each outcome, the 'random variable' gives us a real number.
4. Begin with the following lineup: $n!$, n^{10}, 2^n. Consider $n^{\log(n)}$ versus $\log(n)^n$.

Exercise 7.1

1. A point on the first line is represented by the position vector

$$\mathbf{l}(\lambda) = (4, 5, 1)^T + \lambda (1, 1, 1)^T = (4 + \lambda, 5 + \lambda, 1 + \lambda)^T$$

for some real number λ. Similarly, a point on the second line has position vector

$$\mathbf{m}(\mu) = (5, -4, 0)^T + \mu (2, -3, 1)^T = (5 + 2\mu, -4 - 3\mu, \mu)^T$$

for some real μ. For the lines to intersect, there must exist values of λ and μ for which the two position vectors are the same, namely

$$\mathbf{l}(\lambda) = \mathbf{m}(\mu). \tag{10.3}$$

The above vector equation corresponds to three scalar equations:

$$4 + \lambda = 5 + 2\mu \tag{10.4}$$

$$5 + \lambda = -4 - 3\mu \tag{10.5}$$

$$1 + \lambda = \mu. \tag{10.6}$$

Eliminating μ from Eqs. (10.4) and (10.6), we obtain

$$4 + \lambda = 5 + 2(1 + \lambda)$$

which yields the solution $\lambda = -3$, $\mu = -2$. This solution also satisfies Eq. (10.5), and hence it satisfies all three equations. Substituting these values in Eq. (10.3) we obtain

$$\mathbf{l}(-3) = \mathbf{m}(-2) = (1, 2, -2)^T$$

which is the position vector of the common point of the two lines. □

3. Assign symbols to the line and the parabola. Use the same symbols in the corresponding equations.

Exercise 7.2

(a) The statement of the theorem is imprecise in several respects.

The nature of the numbers x and y is not specified (the inequality would be meaningless for complex numbers).

The case in which one of x or y is zero should be excluded, since in this case the left-hand side of the inequality is undefined.

The statement is false unless the inequality is made non-strict; indeed the equality holds for infinitely many values of x and y.

There are several flaws in the proof.

The basic deduction is carried out in the wrong direction, which proves nothing. (Proving that $P \Rightarrow$ TRUE gives no information about P.)
The assertion 'the last equation is trivially true' is, in fact, false for $x = y$.
The writing is inadequate, without sufficient explanations, and also imprecise (the expression $(x - y)^2 > 0$ is an inequality, not an equation).

(b) Revised statement:

THEOREM: *For all nonzero real numbers x and y, the following holds:*
$$\frac{x^2 + y^2}{|xy|} \geq 2.$$

PROOF: Let x and y be real numbers, with $xy \neq 0$. We shall deduce our result from the inequality $(x \pm y)^2 \geq 0$. We begin with the chain of implications:
$$(x \pm y)^2 \geq 0 \implies x^2 \pm 2xy + y^2 \geq 0 \implies x^2 + y^2 \geq \mp 2xy.$$

Now, since xy is non-zero, we divide both sides of the rightmost inequality by $|xy|$, to obtain
$$\frac{x^2 + y^2}{|xy|} \geq \mp 2 \frac{xy}{|xy|}.$$

The expression $xy/|xy|$ is equal to 1 or -1, and by choosing an appropriate sign we can ensure that $\pm xy/|xy| = 1$. Our proof is complete. $\square$

Exercise 7.3

2. The flaw is subtle: it's related to the incorrect inversion of the logarithmic function. In Sect. 7.7.3 we dealt with the same problem in a simpler setting.

Exercise 7.4

1. Let X be a compact set, and let f be a continuous function over X. RTP: f is uniformly continuous.
3. Let $f(x)$ be a polynomial with integer coefficients, and assume that f is not a constant. RTP: There is an integer n such that $f(n)$ is not prime.
5. Let D be the interior of the unit disc, and let f be a bi-unique analytic mapping of D into itself. RTP: f is a linear fractional transformation.
7. Let P be a Pisot substitution, let M be its incidence matrix, and let f be the characteristic polynomial of M. RTP: f is irreducible over $\mathbb{Q}$.
9. Let D be a differential of the second or third kind. RTP: There is a normal differential N such that $D - N$ is a differential of the first kind.

Exercise 7.5 Consider Euclid's algorithm.

Exercise 8.1 The difficulty in a computer-assisted proof consists in converting the two sides of the identities (8.4) to the same form. In the following Maple codes the function expand is used for this purpose. In each case, the output of the last expression is the boolean constant TRUE.

```
1. a:=k->2*k-1:
   F:=n->n^2:
   a(1)=F(1) and F(k)+a(k+1)=expand(F(k+1));
```

3. The function sum performs symbolic summation.

```
a:=k->k^3:
F:=n->(sum(j,j=1..n))^2:
a(1)=F(1) and expand(F(k)+a(k+1)-F(k+1))=0;
```

5. The base case is $n = 2$.

Exercise 8.4

1. The formula is given in Sect. 2.3.1.
3. Define $f^n = \underbrace{f \circ f \circ \cdots \circ f}_{n}$, and $x_n = f^n(x)$, with $x_0 = x$. Then

$$(f^n(x))' = \prod_{k=0}^{n-1} f'(x_k).$$

Exercise 8.5

1. The base case is not proved.

Exercise 9.1

1. Divide the square into four suitable regions.
2. Consider the function giving the number of friends of a person.
3. The boxes are the elements of the set; what are the objects?

Exercise 9.2

1. The denominator of $f(x)$ vanishes at $x = (-1 \pm \sqrt{5})/2$. Hence the domain of the function should be restricted to $\mathbb{R} \setminus \{(-1 \pm \sqrt{5})/2\}$.
3. The value of the function at a/b is not well-defined, because a and b are not necessarily co-prime. Such a value should be changed to $|a - b|/\gcd(a, b)$. Alternatively, one could change the domain to the set of reduced rationals.

Exercise 9.3

1. FAULTS:
   The integer m must be positive, otherwise the summation is undefined. Alternatively, the sum should be defined to be 0 if $m < 1$.
   The set B is undefined. In particular, we cannot conclude that the reciprocal of each term of the sequence is well-defined.

The co-domain of f is not necessarily B, even if the sum is well-defined.
The symbol a is introduced but not used.
To improve clarity, the dependence of f on the sequence should be made explicit, as a parameter or a second argument. For coherence, this quantity should be called b, not a. Stating that the sequence is infinite would also be helpful.

REVISED STATEMENT:
Given any infinite sequence $b = (b_1, b_2, \ldots)$ of non-zero complex numbers, we define the function f_b as follows:

$$f_b : \mathbb{N} \to \mathbb{C} \qquad m \mapsto \sum_{k=1}^{m} b_k^{-1}.$$

(The set $\mathbb{C}$ could be replaced by any field.)

Exercise 9.4

1. Consider the number of twin primes smaller than a given bound.
2. Is the representation of an even integer as a sum of two primes unique?
3. Consider the smallest non-negative integer t for which the sequence with initial condition $x_0 = n$ has $x_t = 1$.

Exercise 10.1

1. The writing is clumsy; remove reference to project and author in the first sentence; the acronym CF is not used; the verb 'examine' is over-used.
2. Bad first sentence: 'an example of this is that'; the numeral '2' is inappropriate; the main topic (accelerator modes) should be mentioned in the opening sentence; a symbol is introduced but not used; the last two sentences could be merged into one.
3. Inappropriate and heavy use of symbols obscures the main message; introduce a concise symbol for $GQ(s, s)$; the conjecture is formulated twice; the last sentence is clumsy.
4. The first sentence should be re-arranged; some symbols are unnecessary, e.g., those specifying the range of k-values.

Exercise 10.3 How does k depend on n when n is large? Consider Stirling's formula[1] for the factorial.

Exercise 10.4

1. The following Maple code generates the sequence (a_n):

```
N:=10000:
a:=array(1..N,[1]):
for n from 2 to N do
```

---
[1] James Stirling (Scottish: 1692–1770).

```
        if isprime(n^2+n+41) then
          a[n]:=a[n-1]+1
        else
          a[n]:=a[n-1]
        end if
      end do:
```

We start with the lower bound $f(n) = \sqrt{n}$, while the counterexample (7.8) shows that $g(n) \leqslant 40n/41$ for $n \geqslant 41$. The following code displays the ratios $f(n)/a_n$ and $g(n)/a_n$ within the same plot:

```
with(plots):
f,g:=n->sqrt(n),n->n*40/41:
display(plot([seq([n,f(n)/a[n]],n=1..N)]),
        plot([seq([n,g(n)/a[n]],n=1..N)],color=blue));
```

Sharpen these bounds by experimenting with fractional powers and logarithms. (In particular, choose g so that $g(n)/n \to 0$.) A pertinent reference is Golbach's theorem (1752), which states that no non-constant polynomial with integer coefficient can be prime for all (or all sufficiently large) values of the indeterminate, see [15, Theorem 21].

2. This code generates and plots the sequence (z_n):

```
a:=n->n^2/50:
N:=5000:
TwoPiI:=evalf(2*Pi*I):
z:=array(1..N):
z[1]:=evalf(exp(TwoPiI*a(1))):
for n from 2 to N do
  z[n]:=z[n-1]+evalf(exp(TwoPiI*a(n))):
end do:
plot([seq([Re(z[n]),Im(z[n])],n=1..N)]):
```

To increase speed, the exponentials are converted to floating-point with `evalf`. These sequences produce beautiful patterns, but the latter will become apparent only if several thousand terms are plotted.

Gauss first studied these sums in the case $a_n = n^2/p$, $N = p$, with p prime. This is a **quadratic Gauss' sum**, which can be the starting point for a bibliographical search.

3. (a) If $a_n \in \mathbb{N} \setminus A$, then q_n is divisible by 2, 3, or 5.
   (b-c) The following code generates the sequence (a_n):

```
N:=500:
a:=[seq(nextprime(10^(n-1))-10^(n-1),n=1..N)]:
```

This computation is very time-consuming, as it deals with huge integers. Define the sequences of sets

$$A_n = \{a_k : k \leqslant n\}, \qquad W_n = A \setminus A_n, \qquad n \geqslant 1,$$

where A is given in (10.2). Let a_n^+ be the largest element of A_n and let a_n^- be the smallest element of W_n. Both sequences are non-decreasing. The set $f(\mathbb{N})$ is unbounded if and only if $a_n^+ \to \infty$, whereas the condition $a_n^- \to \infty$ is seen to be equivalent to $f(\mathbb{N}) = A$.
The following code generates the sequences $(a_n^\pm)$:

```
A:={$1..N} minus
    select(n->igcd(n,10)>1 or (n mod 3)=2, {$1..N}):
ap,am:=array(1..N),array(1..N):
for n to N do
    {op(a[1..n])}:
    A minus %:
    ap[n]:=max(%%):
    am[n]:=min(%%)
end do:
```

You may consider replacing a_n^+ with the less volatile arithmetical mean of the elements of A_n, generated with the command `add(k,k=a[1..n])/n`. A plot of few hundred terms of this sequence should convince you that $f(\mathbb{N})$ is unbounded, but you must decide if your data provide enough support to the conjecture that $f(\mathbb{N}) = A$.
The bibliography could refer to expository works on the randomness of prime numbers, e.g., [38, Chap. 3].

References

1. Alcock L., Simpson, A.: Ideas from mathematics education: an introduction for mathematicians, The Higher Education Academy, Maths, Stats and OR Network, School of Mathematics, The University of Birmingham, Birmingham UK (2009). ISBN: 978-0-9555914-3-3
2. Ahlfors, L.V.: Complex Analysis, 3rd edn. McGraw-Hill, New York (1979)
3. Anagnostopoulou, V.: Sturmian measures and stochastic dominance in ergodic optimisation. PhD Thesis, Queen Mary, University of London (2008)
4. Berkovich, V.G.: Spectral Theory and Analytic Geometry Over Non-archimedean Fields, vol. 33. Mathematical Surveys and Monographs. Americal Mathematical Society, Providence (1990)
5. Borwein, P.: Computational Excursions in Analysis and Number Theory. Springer, New York (2002)
6. Caldwell, C.: The Prime Pages. The University of Tennessee at Martin. http://primes.utm.edu/ (2014)
7. Carlitz, L.: Finite sums and interpolation formulas over $GF[p^n, x]$. Duke Math. J. **15**, 1001–1012 (1948)
8. Cvitanović, P., Artuso, R., Mainieri, R., Tanner, G., Vattay, G.: Chaos: Classical and Quantum, `ChaosBook.org`. Niels Bohr Institute, Copenhagen (2012)
9. Cohn, H.: Advanced Number Theory. Dover Publications, New York (1980) (Originally published as A second course in number theory, Wiley, New York (1962))
10. Denisov, D., Wachtel, V.: Conditional limit theorems for ordered random walks, Electron. J. Probab. **15**, 292–322 (2010) (MR2609589)
11. Dunham, W.: Euler, The Master of Us All, vol. 22. The Dolciani Mathematical Expositions. The Mathematical Association of America, Washington, DC (1999)
12. Edwards, H.M.: Fermat's Last Theorem. Springer, New York (1977)
13. Gillman, L.: Writing Mathematics Well. The Mathematical Association of America, Washington, DC (1987)
14. Gouvêa, F.Q.: p-adic Numbers, 2nd edn. Springer, Berlin (2000)
15. Hardy, G.H., Wright, E.M.: An Introduction to the Theory of Numbers, 5th edn. Oxford University Press, Oxford (1979)
16. Harriss, E.O.: On canonical substitution tilings. PhD thesis, Imperial College, London (2004)
17. Hasse, H.: Number Theory. Springer, Berlin (2002). (reprint of the 1980 edn)
18. Hensel, K.: Theorie der Algebraishen Zahlen. Teubner, Leipzig (1908)
19. Higham, N.J.: Handbook of Writing for the Mathematical Sciences, 2nd edn. SIAM, Philadelphia (1998)
20. Houston, K.: How to Think Like a Mathematician. Cambridge University Press, Cambridge (2009)

21. Jung, W.: Core entropy and biaccessibility of quadratic polynomials, preprint (2014). arXiv:1401.4792v1
22. Krantz, S.G.: A Primer of Mathematical Writing. American Mathematical Society, Providence (1997)
23. Knuth, D.E.: The TeXbook. Addison-Wesley, Reading (1984)
24. Knuth, D.E., Larrabee, T.L., Roberts, P.M.: Mathematical Writing. The Mathematical Association of America, Washington, DC (1989)
25. Lamport, L.: LaTeX: A Document Preparation System, 2nd edn. Addison-Wesley, Reading (1994)
26. Lange, S.: Undergraduate Analysis, 2nd edn. Springer, New York (1997)
27. Text processing using LATEX, Department of Engineering, University of Cambridge http://www.eng.cam.ac.uk/help/tpl/textprocessing/ (2014)
28. Li, H.C.: p-adic dynamical systems and formal groups, Compos. Math. **104**, 41–54 (1996)
29. Li, H.C.: When is a p-adic power series an endomorphism of a formal group? Proc. Amer. Math. Soc. **124**, 2325–2329 (1996)
30. Li, H.C.: Isogenies between dynamics of formal groups. J. Number Theor. **62**, 284–297 (1997)
31. Liebeck, M.: A Concise Introduction to Pure Mathematics. Chapman and Hall, Boca Raton (2000)
32. Morton, P., Silverman, J.H.: Periodic points, multiplicities, and dynamical units. J. Reine Angew. Math. **461**, 81–122 (1995)
33. Narkiewicz, W.: Finite polynomial orbits. A survey. In: Halter-Koch, F., Tichy, R.F. (eds.) Algebraic Number Theory and Diophantine Analysis. Proceeding of the International Conference, pp. 331–338, Graz, Austria, Aug 30–Sept 5 1998, Walter de Gruyter, Berlin (2000)
34. Concise Oxford English Dictionary, 12th edn. Oxford University Press, Oxford (2011)
35. Pichler, R.H.: Calculating the order of a matrix. MSc thesis, Queen Mary, University of London (1992)
36. Suppes, P.: Introduction to Logic. Dover Publications, New York (1999) (Originally published by D. Van Nostrand Company, New York (1957))
37. Strunk, W., White, E.B.: The Elements of Style, 4th edn. Longman, New York (2000)
38. Tenenbaum G., Mendès France, M.: The Prime Numbers and their Distribution, vol. 6. Student Mathematical Library. American Mathematical Society, Providence (2000)
39. Torii, T.: One-dimensional formal group law of height n and $n-1$. PhD Thesis, Johns Hopkins University (2001)
40. Trask, R.L.: The Penguin Guide to Punctuation. Penguin Books Ltd., London (1997)
41. Truss, L.: Eats, Shoots and Leaves: the Zero-Tolerance Approach to Punctuation. Profile Books, London (2003)
42. Vivaldi, F.: Experimental Mathematics with Maple. Chapman and Hall, Boca Raton (2001)
43. Wirsching, G.J.: The Dynamical System Generated by the $3x+1$ Function. Lecture Notes in Mathematics, vol. 1681. Springer, Berlin (1998)
44. Davis, S., Swinburne, D., Williams, G. (eds.).: Writing matters: the Royal Literary Fund report on student writing in higher education. The Royal Literary Fund, London (2006) (Electronic copy available from http://www.rlf.org.uk/fellowshipscheme/research.cfm (2014))

Index

Symbols
$3x+1$ conjecture, 160, 164
[♯], 20, 25, 40, 45, 73, 74, 86, 90, 109–111

A
Abscissa, 18
Affine, 85
AGM inequality, 139
Algebraic product, 26, 27
Algebraic sum, 26, 27, 49
Alphabet, 96
Ambient set, 12, 14, 41, 70, 75
Antecedent, 54
Archimedes
  Archimedean property, 64
  constant, 44, 98
Argand plane, *see* complex plane
Argument, 22, 94
Arithmetic progression, 104
Ascending, 47
Assignment statement, 10, 16
Associative, 12, 157
Axiom, 113
  subset, 15

B
Base, 15
Bernoulli inequality, 139, 145
Bertrand's postulate, 158
Bi-unique correspondence, 28, 57
Bijective, 23
Binary, 12, 52, 95
Binomial, 33
  coefficient, 45
  theorem, 139

Boolean function, 95
Boolean operator
  AND ($\wedge$), 53
  NOT ($\neg$), 53
  OR ($\vee$), 54
  XOR ($\veebar$), 54
  if ($\Leftarrow$), 55
  if and only if ($\Leftrightarrow$), 55
  only if ($\Rightarrow$), 54, 116
Both ends method, 124, 147, 153
Bound, 154
  lower, 81
  strict, 141
  upper, 81, 146
Bounded
  function, 81
  set, 80
Bounded away from zero, 81

C
Cantor, 9
Cardinality, 11, 28
Cartesian plane, 18, 108
Cartesian product, 12, 13, 18, 20, 22, 48, 61, 95, 98, 108
Cauchy-Schwarz inequality, 139
Ceiling, 86
Chain rule of differentiation, 130
Characteristic function, 57, 58, 63, 95
Circle, 18, 29, 95
Closure, 83, 101
Co-domain, 21, 158
Co-prime, 16
Coefficient, 33
Commutative, 12, 39
Complement, 12

Complex
    number, 17, 98
    plane, 17
Component, 94
Composition of functions, 23, 33, 44, 49
Computer-assisted proof, 144
Conclusion, 54
Conductor, 93
Congruence
    class, 27
    operator, 52
Congruent, 27
Conjecture, 125, 160, 164
Conjunction, 54, 121, 155
    conjuncts, 121
Consequent, 54
Continuous, 83
Contradiction, *see* proof
Contrapositive, 56, 119, 120
Convergence, 88
Converse, 55, 56, 129
Coset, 123
Countable, 28
Counterexample, 124

D

De Morgan, 140
    laws, 56, 64, 68
Decimal point, 15, 134
Dedekind, 99, 143
Definition, 10, 157
    recursive, 107
Degree, 33
    total, 33
Descending, 47
Difference, 15
Dimension, 32
Dirichlet's box principle, 153, 164
Disc, 18
Discontinuous, 84, 86
Disjoint, 11
Divergence, 88
Divide and conquer method, 103, 104
Divisibility, 16
    operator, 53
Domain, 21, 158
Domain name, 176
Double sum, 35
Dummy variable, 34

E

e, *see* Napier's constant

Eigenvalue, 93
Element, 9, 31, 94
Ellipsis, 13, 14, 19, 41
    raised, 16, 19, 33, 34
Empty set, 10, 13, 157
Entry, 94
Envelope, 101
Equality, 52
Equation, 37, 58
    algebraic, 44
    cartesian, 29, 101
    differential, 38, 49
    functional, 40, 49
    polynomial, 142
    simultaneous, 39
    transcendental, 44
Equivalence class, 70
Equivalence relation, 70
Equivalent
    equations, 50
    sets, 28, 29
    statements, 56
Estimate, *see* bound
Euclid, 126, 146
    algorithm, 191
    theorem, 123, 147, 151
Euler, 99, 125, 142, 180
    φ-function, 35
    gamma-function, 98
Euler-Mascheroni constant, 98
Eventually constant, 87
Eventually periodic, 87, 133, 160, 171
Exceptional set, 93
Existence proof, 151
    constructive, 151
    effective, 154
    non-constructive, 151, 152
Existence statement, 66, 151
Exponent, exponentiation, 15
Expression, 40
    algebraic, 43
    analytical, 44
    arithmetical, 43
    combinatorial, 45
    compound, 54
    integral, 45
    logical, 53
    nested, 44
    relational, 51, 52, 70
    trigonometric, 44

F

Factorial, 45, 60, 159

Index 201

Farey sequence, 48
Fermat, 142, 156
Fibonacci sequence, 159
Finite, 11
Floor, 86
Fractional part, 86
Function, 21, 22, 32, 98
    affine, 85
    algebraic, 44
    arithmetical, 87
    boolean, 57
    bounded, 81
    circular, 44
    constant, 23
    continuous, 83, 84
    decreasing, 78
    differentiable, 84
    discontinuous, 84
    even, 78, 154
    increasing, 78
    invertible, 24
    linear, 85
    monotonic, 78
    negative, 77
    odd, 78
    periodic, 49, 79
    positive, 77
    regular, 84
    singular, 84
    smooth, 84
    transcendental, 44
    trigonometric, 44
Functional, 95
Fundamental period, 79
Fundamental theorem of arithmetic, 155

G
Gauss, 27, 169, 194
Gcd, *see* greatest common divisor
General term, 31, 32, 34, 100
Geometric progression, 35
Goldbach
    conjecture, 126, 164
    theorem, 194
Graph, 22, 94, 98, 110
Greatest common divisor, 16, 22, 27, 41
Group, 38, 98, 123
    axioms, 114, 156
    identity, 156

H
Hcf, *see* greatest common divisor

Hypothesis, 54

I
i, *see* imaginary unit
Identity, 38–40, 59, 67, 133
Identity function, 22, 49
If, iff, *see* boolean operator
Ill-defined, 157
Image, 23, 68, 81
Imaginary
    part, 17
    unit, 17, 98
Implication, *see* boolean operator, only if
Inclusion operator, 52
Incremental ratio, 84
Indeterminate, 33, 94, 161
Indeterminate equation, *see* identity
Induction, 140, 158
    basis of, 143
    inductive step, 143
    principle, 143
    strong, 140, 146, 159
Inequality, 52
Infinite, 11
Infinite descent, 140, 142
Infinity, 18
Initial condition, 159
Injective, 23, 64
Integer, 13, 97
Integral equation, 49
Intermediate value theorem, 154
Intersection, 11, 20
    empty, 11, 157
Interval, 58
    closed, 17
    half-open, 17, 29
    open, 17
Inverse, 23
Inverse image, 24, 57
Invertible, 24
Involution, 24, 49
ISBN, 175
Isomorphism operator, 52
ISSN, 175
Italic fonts, 98

K
Kroneker, 99

L
Lambert, 99

LaTeX, 177
Latin
    Q.E.D., 114
    ad infinitum, 142
    definiendum, 10
    definiens, 10
    e.g., 1
    et al., 176
    etc., 1
    i.e., 109
    non sequitur, 128
Lcm, *see* least common multiple
Least common multiple, 16
Least counterexample, 140
Lemma, 113
Length, 31
Limit, 83, 84, 88
Linear, 85
Local property, 81
Logical
    constant, 51
    equivalence, 56
    expression, 53
    operator, 51, 53
    tag, 114
Lower limit, 34

M
Map, mapping, 21, 95
Mathematical induction, *see* induction
Maximum, 81
    local, 82
Member, 11, 94
Membership operator, 52
Metric space, 94
Micro-essay, 108, 109, 111, 170
Micro-project, 179, 193
Minimal period, 79
Minimum, 82
    local, 82
Minkowski, 26
    sum (product), *see* algebraic sum (product)
Modular arithmetic, 27, 39, 162
Modulus, 27
Monomial, 33
Monotonic, 78
Multiplicity, 10
Multiset, 10, 156
Multivariate, 33

N
Napier's constant, 44, 98
Natural number, 13
Necessary and sufficient, 122
Negation of logical expression, 53, 68
Neighbourhood, 81
    of infinity, 82, 87
Nested
    expression, 42, 44, 69
    parentheses, 14
    sequence of sets, 47, 48, 87
Number
    complex, 17, 98
    integer, 13
    natural, 13
    rational, 14, 97
    real, 17, 97

O
One-parameter family, 95, 101
One-to-one, *see* injective
Onto, *see* surjective
Operand, 12
Operator, 21, 62, 95
    arithmetical, 95
    assignment, 10, 16, 36
    binary, 52, 53, 95
    boolean, 12
    differentiation, 95
    divisibility, 52, 60
    equivalence, 55
    inclusion, 60
    isomorphism, 52
    logical, 51, 53, 95
    membership, 60
    orthogonality, 52
    relational, 51, 52, 70, 95
    unary, 53, 95
Opposite, 15
Order, 159
Ordered pair, 12, 32
Ordering, 78
    partial, 71
Ordinate, 18
Orthogonality operator, 52

P
Pairwise disjoint, 11
Parameter, 95, 101, 161, 171–173, 189
Partition, 26, 27, 70
Peano, 143

axioms, 140
Period, 79
Periodic, 79
Pi, π, *see* Archimedes' constant
Piecewise, 85, 86, 116
Pigeon-hole principle, *see* Dirichlet's box principle
Point, 94
    boundary, 82
    interior, 82
    isolated, 82
Polynomial, 33, 44
    homogeneous, 33
    multivariate, 33
    univariate, 33
Postulate, *see* axiom
Power series, 37, 104, 105
Power set, 26, 29, 49, 95
Predicate, 15, 51, 57, 58, 61, 63, 95, 105, 155
Prime, 16, 97, 109, 123, 126, 127, 151, 155
Primitive, 93
Product, 15
    infinite, 37
    symbol, 34
Product of sets, *see* algebraic product
Proof, 113
    both ends method, 124, 147, 153
    by cases, 116, 152
    by contradiction, 116, 123, 129, 140, 142, 143, 153
    by contrapositive, 119, 120
    by induction, 139
    direct, 118
    of conjunction, 121
Proper
    divisor, 16
    subset, 11
Proposition, 113
Punctuation, 3, 102

Q

Quadratic
    function, 85
    Gauss' sum, 194
    irrational, surd, 43
    polynomial, 33, 101, 126
Quadruple, 32
Quantifier, 51
    existential, 59, 62, 63, 155
    universal, 59, 62, 63, 118
Quotient, 15
Quotient set, 70

R

Range, *see* image
Rational
    expression, 43
    function, 33, 43, 44
    number, 14, 97
    point, 18
Ray, 18
Real
    function, 77
    line, 17
    number, 17, 97
    part, 17
    sequence, 77
Reciprocal, 15, 24, 42, 81
Recursive
    definition, 107, 158
    sequence, 107, 159
Reduced form, 14
Regular, 84
Relation, 70
    anti-symmetric, 71
    equivalence, 70
    reflexive, 70
    symmetric, 70
    transitive, 70
Relational
    expression, 51, 52, 70
    operator, 51, 52, 70
Representation of sets, 19, 28, 29, 48
Restriction, 23, 24
Riemann
    hypothesis, 126
    zeta-function, 99
Roman fonts, 98
Root of polynomial, 38, 85, 154
Russell-Zermelo paradox, 14

S

Sequence, 31, 99
    doubly-infinite, 32, 160
    finite, 31, 32, 48
    infinite, 31, 48
    invertible, 160
    of sets, 47
    periodic, 160
    recursive, 159
Series, 36
    convergent, 36
    divergent, 36
    geometric, 134
    sum of, 36
Set, 9, 41, 96

abstract, 28, 83
bounded, 80
closed, 82
finite, 11
infinite, 11
open, 82
ordered, 71
partially ordered, 71
Set difference, 11, 156
Set equation, 49, 50
Set operator, 12, 95
Sigma-notation, 34
standard form, 35, 59
Sign function, 79
Singular, singularity, 84
Solution set, 38
Space, 94
Square root, 43, 115, 130
Standard definition, 13
Step function, 86
Stirling, 193
Sub-sequence, 32, 99
Subscript, 31, 32, 97
Subset, 11, 60
operator, *see* inclusion operator
Sum, 15
Sum of sets, *see* algebraic sum
Sum-free set, 49
Summation, 34
index, 34
limits, 34
range, 34
symbol, 34
Surjective, 23, 64
Symmetric difference, 11
System of equations, 39

T
Tautology, 56
Term, 31, 94
TEX, 177
Theorem, 52, 113
conditional, 126
Thesis, 165
Transitive, 115
Triangular number, 106
Triple, 22, 23, 30, 32
Trivial, 133
equivalence, 70
subset, 11
Truth table, 53, 55

Twin-primes conjecture, 126, 127, 164

U
Unary, 53, 95
Uncountable, 28
Union, 11
Unique existence, 155
Unit
circle, 18
cube, 18
disc, 18
hypercube, 18
interval, 18
sphere, 19
square, 18
Univariate, 33
Unknown, 37, 94, 98
Upper limit, 34
URL, 176

V
Value
of a function, 21, 23
of an expression, 40
Vanishing, 38
identically, 38
Variable, 22, 94
dummy, 34
random, 111
Vector, 32, 98–100
dimension, 32
Vector space, 94

W
Weierstrass, 99
Well-
behaved, 84
defined, 157
ordered, 71
ordering principle, 140
Wilson's theorem, 122
Without loss of generality, 80, 135, 163

Z
Zermelo definition, 14, 38, 57, 63, 103
Zero
function, 38
of equation, 38

Printed in Great Britain
by Amazon